W9-BWW-814

Guided Engineering Design PROJECT BOOK

By
CHARLES E. WALES
West Virginia University

ROBERT A. STAGER
University of Windsor

THOMAS R. LONG
West Virginia University

WEST PUBLISHING CO. ST. PAUL, NEW YORK, LOS ANGELES
SAN FRANCISCO, BOSTON

COPYRIGHT © 1974 By WEST PUBLISHING CO.
All rights reserved
ISBN: 0–8299–0002–0
Printed in the United States of America

PREFACE

Guided Design is a new approach to teaching and learning which focuses on developing the student's decision-making skills as well as teaching specific concepts and principles. To accomplish both ends this package contains two books:

> Guided Engineering Design presents design material which includes an introduction to the decision-making process and the "Instruction-Feedback" pages for four design projects.

> An Introduction to Engineering Calculations: The Energy Balance, Mechanical Advantage, Fluid Flow and Static Forces presents the subject matter concepts pertinent to each of the four designs.

This material is appropriate for a first course in engineering design at the freshman or sophomore level; only algebra and trigonometry are required as prerequisites.

In a Guided Design class the students work in small groups to solve meaningful open-ended problems which require them to think logically, gather information, communicate ideas and use each of the decision-making steps. The students are guided through the solution of each design problem by a series of printed "Instruction-Feedback" pages, by their discussion with other students in their design team, and by the teacher, who acts as a consultant. The students do the thinking, they must make value judgments, and they play the role of the professional decision-maker.

In the Guided Design systems approach each open-ended problem establishes a need for a unit of subject matter, which each student is expected to learn on his own outside of class, in this case by using the companion programmed text. This approach should help the student learn that facts, concepts and principles are part of the background information required for the decision-making process. This organization should establish a pattern which will serve him well after he leaves school-- where continued independent learning is a prerequisite to success. In this engineering course the student is expected to gather the information he needs from the self-study materials, from the library, or through experimental work. While the student is learning to be independent, some class time may be required to check on his progress, examine his homework for errors, provide help, and give examinations. However, even during this period, most of the class time can be freed for decision-making activities associated with the design projects.

In the Guided Design system the prime role of the teacher is that of guide, prompter, manager and consultant. During class the teacher should plan to move from group to group listening, asking leading questions and encouraging the students to participate in the decision-making process. He should also exert some control over

620.8042018
W172g
v.2
c.2

e flow of the project Instruction-Feedback pages, because students sometimes have a tendency to try to shortcut the decision-making process and ask for feedback before they have achieved an appropriate level of thinking.

Guided Engineering Design was written to provide several options to the teacher who uses it. At West Virginia University the book is used in a two-semester sequence of 3-credit engineering design classes. However, we recognize that many of those who teach a course such as this may want to add other design work and subject matter material which suits their own situation. To provide for this the projects in Guided Engineering Design involve either four or five weeks of work, so the complete book includes enough material for most, but not all of the time available in a two-semester or three-quarter course. Those who teach a one-semester course have the option of using two of the projects from the book and adding some of their own work or using three of our four projects. Those who teach a one-quarter course can use one or two of the four projects.

However it is used, the point of this book is to help each student learn how to attack an open-ended design problem--using Instruction-Feedback pages to model the decision-making process in slow motion--and to help the student see that the engineering-science concepts he learns can help him make a better decision. Thus, this book provides the first step in the development of the student as an independent, creative, adaptive engineer. It prepares the student for later courses where he should work with case studies, simulations, games, and finally some form of ternship-type experience.

Neither this book nor the concept of Guided Design could have been developed without the support provided by Dean Chester A. Arents and many of the engineering faculty at West Virginia University--we do appreciate this support. The Exxon Education Foundation also played an important part by providing a grant for the initial development work. The book was written and tested over a four-year period in classes at WVU and many faculty and students made important contributions to its development. Others who have used this book and provided valuable feedback are Richard Graham, Wichita State University; Robert Thornhill, Wayne State University, Herbert Hartman, Potomac State College, and Kendall Hall at Parkersburg Area Community College. In addition, we would like to thank Peter A. Popovich, a student at WVU, who helped write the material on the slide rule. And, of course, we are most indebted to our secretary, Janet Oberlechner, who has typed and retyped this material so often she must know it all by now.

Many faculty have shown an interest in the concept of Guided Design and it has already been successfully applied in a variety of educational settings: a high school course on the environment; college course work in several engineering disciplines, chemistry, forestry, drama, education, and a multidisciplinary liberal

arts course; all the professional undergraduate course work in a chemical engineering program; four courses in the first semester of a graduate program in rehabilitation counseling; in an industrial training program; and in workshops on educational
systems design for faculty. In 1973, the Exxon Education Foundation began a new
program to encourage faculty in all disciplines to consider the adoption of a few
selected educational innovations including the Guided Design systems approach. We
hope this program will strengthen both your interest in this book and the potential
application of this approach in other courses in your program.

GUIDED ENGINEERING DESIGN

TABLE OF CONTENTS

INTRODUCTION

This book, <u>Guided Engineering Design</u>, is based on a new concept of course design and operation called Guided Design. The Guided Design process focuses on developing both your decision-making skills and your ability to handle new principles and concepts. The need for a student to learn pertinent principles and concepts should be obvious, the better command he has of basic ideas, the better prepared he will be to use these as a professional. But what about decision-making, is this also important? Many people in business, industry, and government claim that competent decision-making is the basis for most of the progress we have made during the past one hundred years. Some psychologists claim that the ability to make a decision and act on it is a key component of full psychological health. If these viewpoints are correct, all of us should be concerned with developing each student's decision-making ability.

The Guided Design approach is based on the belief that the <u>educational system</u> <u>should provide experiences which make it possible for each student to learn and</u> <u>demonstrate that:</u>

Goal 1--Knowledge

He can recall, manipulate, translate, interpret, predict and choose appropriate facts, concepts and principles as he solves single-answer problems.

Goal 2--Open-Ended Problems

He can learn by himself, think for himself, think logically, gather and organize the information required to make decisions, communicate ideas and use the decision-making process including analysis, synthesis and evaluation in the solution of open-ended problems.

Goal 3--Values

He can develop and make use of his own value system, which involves a concern for people, the environment and economics.

Goal 4--Professional Experience

He can operate as a professional in the discipline he is studying; i.e., he can work both independently and as part of a team to solve open-ended problems creatively and humanely.

To help you achieve Goal 1, the Knowledge goal, we have provided a companion programmed text, An Introduction to Engineering Calculations. The Guided Engineering Design book you are now reading is designed to help you achieve the other three goals. In this book we model the decision-making process in slow motion using printed Instruction-Feedback pages. Each Instruction presents some information and asks the students, who are working in small groups, to make a decision about one step in the solution of an open-ended problem. The students do the thinking, but when they complete their work the faculty design consultant, who controls the flow of materials, gives them the printed Feedback so they can compare their decision with that of a professional. As the student's ability develops, the amount of guided support they receive is reduced.

The initial problem in this book is based on the highway sign "Bridge Freezes Before Road Surface." This problem should take about one hour to work through and is intended simply to introduce the steps in the decision-making process. The remainder of the book consists of four design projects. Some of these projects take four weeks to complete, others require five weeks.

Before you begin work on these projects you should realize that it is not necessary for your group to agree exactly with each other, with the printed Instructions or with the Feedback. The point of the group discussion is to stimulate thinking about the point in question. Many of these points have equally acceptable alternate answers. We don't guarantee our answer is the correct one; in fact, your group's answer may be better than ours. However, we do expect that in general you will find our answer both reasonably logical and quite similar to yours. If possible, try to avoid arguments which polarize the group. You are not likely to settle all the issues involved in the time available for your discussion. The importance of this work is the ideas that are being presented and not the particular answer to a given question.

One way to summarize the characteristics of Guided Design is to describe some of its unique operating features. These include the following.

1. The students work on decision-making projects in four to seven-man discussion groups.

2. Different members of each group have an opportunity to act as the student project leader during a given class period.

3. The student project leader is sometimes asked to guide his group through a given step in the decision-making process using the information he receives from the programmed "Feedback."

4. Each "Instruction" asks the group to perform a step in the decision-making process. During the discussion of this step, each student receives feedback on his "thinking" from classmates. If the faculty "consultant" is

available, he may also contribute to the discussion. As soon as a group demonstrates an appropriate depth and direction of thinking to the instructor, they receive feedback on their efforts from the printed material which describes what probably should have been accomplished.

5. All the work the student does is organized around the open-ended problem. The problem establishes a need for the concepts presented in the program.

6. To provide easy access to the program for review, the student completes his own Reference Book, which identifies the major concepts he is expected to learn.

7. The student tests his understanding of new concepts by solving a set of single-answer problems and taking a Self-Quiz at home. Answers are given with the problems. Both the quiz and the problems may be checked in class, where solutions are available for group discussion.

8. The student learns single-answer problem-solving techniques through a series of branched programmed lessons which complement the subject matter programs. Problem solving ability is tested with a problem test.

9. The culmination of each project is a formal report on the group's design, which is prepared by each student or by the group. We recommend that the student or group grade their own report and be encouraged by the instructor's feedback to rewrite to earn this grade if the report falls short of the instructor's expectations.

10. The content-performance objectives given at the end of each program chapter the multiple-choice quiz, the problem quiz and the homework problems form the basis for a comprehensive examination on the new concepts introduced in each design project. We recommend that the student rewrite this examination (or an alternate form) until he demonstrates competence with the concepts and problem solving abilities.

11. Class time is available to discuss not only the design details but also the important moral and philosophical questions which arise during the design project. These discussions may take place within the small groups or in whole class discussions.

We believe that the combination of all these features results in a course which achieves important educational goals and provides a meanginful learning experience for the students who participate in it.

AN INTRODUCTION TO
THE PROCESS OF
DECISION-MAKING

This is a course in decision-making--a basic intellectual process which equips a man to face and solve each of the new problems he meets. Our work begins with a short exercise which involves about one hour's effort. This work will introduce you to the basic concepts of decision-making. The problem we'd like you to consider is one you might discover yourself if you drive in snow country. This discovery would occur when you see the following sign along the highway.

> **BRIDGE FREEZES**
> **BEFORE**
> **ROAD SURFACE**

Your job is to consider the problem suggested by the sign. To do this, we recommend that you form four to six man teams who can discuss the problem and decide what to do. You should take about ten minutes for this stage of your work. Then we will discuss your results. Begin this work now. Do not read further until you have completed your discussion of the problem.

Wales et al.—Engineering Projects Book—1

Even if the people in your group have had some training in decision-making work, they may have used their discussion period to consider the <u>possible solutions</u> to the "Bridge Freezing" problem. Thus, your group may have decided to solve the problem in one of the following ways.

1. Insulate the bridge bottom.
2. Heat the bridge surface
   a. electric, heating wires in the surface.
   b. radiant, heating lamps above the surface.
3. Cover the bridge.
4. Put salt on the bridge surface.
5. Move the bridge to Florida.

Although generating possible solutions such as these is a worthwhile thing to do and a necessary step in the decision-making process, it is not the best step to take at this time. Something else should occur first. In fact, several steps normally precede the possible solutions step. To help you learn what these steps are and an order in which they are likely to occur, we would like you to consider the set of questions given below--questions organized in the form of "branched" programmed instruction.

To properly use this branched program, you should read each question, select an answer from those given in the question, then turn to the numbered section which corresponds to your choice and check your answer. If you select what we consider to be the best answer you will be given further instructions and a new question. If you select one of the other answers, you will be given further information and/or a chance to select another answer from the original set of answers.

	Section 1

If generating possible solutions is not the first step in the decision-making process, then which of the following is the best first step?

Turn to

Section Number	If Your Answer Is
2	Analyze, synthesize and/or evaluate a solution.
3	Identify the specific problem to be solved.
4	State the basic objective or goal of the work.

Turn to the section that corresponds to the answer of your choice.

	Section 2

Your answer is that the first step in this process is to <u>analyze</u>, <u>synthesize</u> and/or <u>evaluate</u> a solution. This is a good choice, but not the best one in this set. The terms analysis, synthesis and evaluation describe three intellectual operations you will perform over and over again. But you do not yet have a basis for performing these operations on any solution in this

problem. Therefore, you should return to Section 1 and make another
choice.

Your answer is that the first step in this process is to <u>identify the specific problem that exists</u>. This answer is correct. To identify the problem you will use the intellectual operations of analysis, synthesis and/or evaluation--these are the thinking processes you must perform to accomplish each step in the decision-making process. However, you cannot begin a solution until you know exactly what the problem is. Thus, the first step is to identify the problem.	Section 3

You may think that the problem you want to solve has already
been identified, but has it? Does the highway sign, "Bridge Freezes
Before Road Surface," state the real problem or does the sign
actually represent just one of the many possible solutions to the
problem?

Turn to

Section Number	If Your Answer Is
5	The sign states the problem.
6	The sign is a solution to the problem.

Read the answer of your choice.

Your answer is that the first step in this process is to <u>state the basic objective or goal</u> of the work. This answer is good, but it is better to do something else before stating the goal. You cannot decide what you want to accomplish through your work until you know what the problem is. Return to Section 1 and make another choice.	Section 4

You believe the sign, "Bridge Freezes Before Road Surface," states the problem. This answer is not correct. Read Section 6 to see why.	Section 5

You think the sign, "Bridge Freezes Before Road Surface," is a solution to the problem. You are correct. This sign was put up by the highway department in an attempt to solve the problem. It is not a very good solution because it does not completely eliminate the problem; it only warns the motorist that he may sometimes have a problem. If you've ever driven from dry pavement onto an icy bridge, you know that it can be quite a problem because you can instantly lose control of your car.	Section 6

If the sign is not the problem, then what is the problem? For

3

example, do we really care if ice forms on the surface of the bridge?

Turn to

Section Number	If Your Answer Is
7	Yes
8	No

Turn to the answer of your choice.

You answered yes, we care about ice forming on the surface of the bridge. Why do we care? Will the ice structurally damage the bridge? No, not if it is just on the surface. It is something that happens because the ice is there that is the real problem and not the ice itself. To attack the real problem, read Section 9.

Section 7

You answered no, we don't care about ice forming on the bridge. That's correct. As long as the ice is only on the bridge surface, it won't structurally damage the bridge so we don't care if it forms there. Thus, it is not the ice itself that is the problem. It is something that happens because the ice is there that is the problem. To attack the real problem, read Section 9.

Section 8

If neither the sign nor the ice is the problem, then what is he problem we must solve through our work? Is it,

Turn to

Section Number	If Your Answer Is
10	to prevent damage to the car?
11	to prevent damage to the bridge?
12	to prevent damage to the people in the car?

Turn to the answer of your choice.

Section 9

This answer is partly correct; the problem is to prevent damage to the car. However, there is a better answer in this set. Go back to Section 9 and choose a different answer.

Section 10

This answer is partly correct. The problem is to prevent damage to the bridge. However, there is a better answer in this set. Go back to Section 9 and choose a different answer.

Section 11

While all the answers given in Section 9 are correct, this is the best answer. The problem is that people are injured when cars skid on the frozen bridge surface. We would like to prevent this injury. If at the same time we can protect both the car and the bridge from damage, so much the better. However, people are our

Section 12

4

first concern.

Identifying the correct problem is an extremely important first step in the solution of any problem. If you have not done this step properly, you are not likely to produce the best solution to the problem because you may miss possible solutions which should not be missed. For example, in the early development of the U. S. space program, a problem similar to the following was posed.

Find a material which will withstand a temperature of $14,000^{\circ}$F for five minutes.

Of course, this problem was related to re-entry of the space capsule and the enormous amount of heat generated during the process.

A great deal of time and money was wasted trying to solve this problem because there was no known material which would withstand the required temperature. Finally, someone realized that they were trying to solve the wrong problem. The new problem was stated as follows.

Find a way to protect the capsule and the man inside during re-entry.

This problem was solved very quickly with the ablation system now in use.

The bridge problem is very similar to this re-entry problem. If you only worry about the ice on the bridge and not the real problem of protecting people, you will miss potentially acceptable solutions. For example, tires which have metal studs and tires with chains solve the people problem, but do not eliminate the ice.

As soon as you have identified the problem that exists, you should be able to state the basic objective or goal of your work. The basic objective or goal is a statement of what you expect to accomplish through your work. This statement will be used to test your final result. Your group should now discuss, agree on and state the goal for this bridge project. Do not read Section 13 until you have stated the goal.

We believe the goal of this bridge project should be to,

> Create a system which will prevent injury to people who
> unexpectedly drive from non-icy pavement onto an icy
> bridge surface. It would also be desirable to prevent
> damage to the car and the bridge.

Note that in this case the goal of the work is essentially a
restatement of the problem that we identified; this is often the
case.

The next step in the decision-making process is to gather
information about the problem. This step may well be repeated
when you perform each of the other steps in the process. In this
case, we must gather information so we can determine the constraints,
assumptions and facts which will affect the work. Constraints are
factors which cannot be changed and limit what you can do. For
example, one constraint might be the money available, another might
be the building materials which exist or the attitude of the people
involved. One assumption which seems reasonable is the impossibil-
ity of eliminating all bridges. A second is the climate; you cannot
change that. One of the facts that will affect your work is the
number of times the bridge freezes each year.

Your group should now discuss these and any other factors which
will affect what they can do. You might also make a list of the
questions which you want answered. Hopefully, these answers will
help you determine additional pertinent constraints, assumptions
and facts. When your list is complete, read Section 14.

Some of the possible constraints, assumptions and facts that
may limit this bridge project are,

1. The money available. (Constraint)

   If little money is available the only solution may be a
   sign to warn the motorist. In fact, this is the solution
   that presently exists.

2. The physical state of the bridge.

   a. Does the bridge exist or is a new bridge to be built?
      (Fact)

   b. If the bridge exists, can traffic be stopped or
      rerouted while changes are made? (Assumption)

   c. What kind of bridge surface exists or is planned?
      Concrete, asphalt, and steel gridwork each present
      different problems. (Fact)

3. How often ice can be expected. (Fact)

   Ice formation once each winter as opposed to once a week
   for 15 weeks will affect the design result.

The answers to some of these questions might actually affect
the goal you established earlier. Other answers will simply affect
your solution.

Up to this point you have identified the problem, established
the goal of the project, gathered information and listed the constraints, assumptions and facts. The next step in the decision-
making process is one of the following.

Turn to

Section Number	If Your Answer Is
15	Generate possible solutions.
16	Analyze possible solutions.
17	Synthesize a detailed solution.

Read the answer of your choice.

Your answer is that we should now begin to generate possible
solutions. This is the proper step to take at this point. And
this is where your creativity, imagination and ability as a dreamer
come to the fore. This is not the time to be critical; it is the
time to stretch all your faculties in a search for potential answers
to the problem. Thus, you should now generate all the solutions
you can and criticize them later. This is, of course, the step you
probably took at the beginning of this bridge project. Now this
step can be based on the correct problem, a clearly stated goal and
an identified set of constraints, assumptions and facts. For

example, if protecting people is the objective and not necessarily removing the ice, you can generate solutions that involve the bridge, the car and the people.

To allow you to proceed with this project, let us assume that we are concerned with a bridge that exists. Also that traffic can be diverted for a few weeks if necessary. The bridge is of reinforced concrete construction. Ice forms at least seven times during the winter, but might occur as often as fifteen times. And we are looking for a low cost solution because all the money spent on this project must come at the expense of other road work, repairs or new construction which might be done. Using this information, you should now generate as many solutions to the problem as you can. Then read Section 18.

Your answer is that we should now begin to <u>analyze possible solutions</u>. This is a good idea, but where do we get the possible solutions? Remember, the list you generated earlier was for the ice problem, not the people problem. Many new solutions may now be available, but they must be generated before you can analyze them. Go back to Section 14 and make a new choice.	Section 16
Your answer is that we should now begin to synthesize a detailed solution to the problem. This must, of course, eventually be done. But you're not ready for this step yet; the act of creating <u>a detailed solution</u> can only occur after you have examined all the alternates and chosen the one that appears to have the greatest potential. Go back to Section 14 and make a new choice.	Section 17
At this point, you have generated a number of potential solutions to the problem. Next you must examine these solutions and attempt to choose the best one or ones to pursue. This step will surely require both analysis and evaluation; it may also include some synthesis. Perhaps the best way to start is to rank the solutions you have generated in order from the best, to the poorest. To rank any set of items you must have some kind of <u>criteria</u> or <u>value system</u>. Thus, at this point you should specify the criteria you will use to rank your possible solutions. The constraints you specified earlier may be of help to you here. When you have finished your list of criteria read Section 19.	Section 18

Three very basic criteria which apply to essentially every project are safety, economics and feasibility. In fact, it is difficult to conceive of a project in which these three criteria are not prime factors.

Now you should use these criteria to rank your possible solutions. One way to do this would be to assign a high - medium - or low cost label to each solution for economics, a good - average - or poor label for the safety factor and a yes - no - or ? for feasibility. Do this job now, then read Section 20.

Your list of ranked solutions may look like the one given below.

Possible Solution	Safety	Economics	Feasible
1. Insulate the bridge bottom.	Good	Low cost	Yes
2. Melt the ice--electricity.	Good	High cost	Yes
3. Cover the bridge.	Average	High cost	Yes
4. Tires, studs, chains.	Good	Low cost	?
5. Melt the ice--salt.	Average	Low cost	Yes
6. Install a warning sign.	Poor	Very low cost	Yes

From the looks of this list it would appear that the best solution is 1, insulating the bridge bottom.

Of course, this simple table does not come anywhere near giving us all the information we need to make a final choice of the best solution. For example, a further analysis would show that we don't know if insulating the bridge bottom will solve the problem; ice may still form. The list also does not include details such as the following.

2. Providing the equipment required to turn on an electric grid at the right time may be an expensive task.

3. Covering the bridge might eliminate some ice, but some- times ice crystallizes directly from the air above the surface. In addition, cars would certainly carry water from the road onto the bridge where it might freeze.

4. Tires with studs and chains are a good solution. How- ever, not everyone will buy or can afford such tires. And studs and chains are hard on the surface of the road when there is no ice or snow.

5. Salt will probably have to be spread by a truck. This means delays because the truck normally would not go to the bridge until the ice had formed.

If you were the engineer on this project, you would consider all these additional factors and still probably choose "insulation" as the first solution to pursue. The next step is to perform a detailed analysis, synthesis and evaluation to determine exactly what might be done and whether or not the solution will satisfy the goal of the project. Let's assume you decide to follow this path. What design variables would you consider and how would you evaluate the results? Would you use theoretical calculations, lab experiments or test an actual bridge? After you have considered all these factors and made your decision, read Section 21.

Section 21

The problem at hand is to prevent the surface of a reinforced concrete bridge from reaching the freezing point of water, $32^{\circ}F$, before the adjacent road surface. Our analysis shows that the road is insulated by the earth below it. And, at least part of the time, the earth can be expected to act as a source of heat which keeps the road surface from freezing. The bridge, however, is exposed to the cold air on both the bottom and the top. When the air temperature drops below $32^{\circ}F$, the bridge surface will cool to that temperature much faster than the road itself. Insulating the bottom of the bridge should help reduce the rate of cooling but the insulation cannot act as a heat source, as does the earth. Thus, we know in advance that insulation will not solve the problem completely.

A detailed analysis can now be made to determine what type of insulation to use and how thick to make it. The answers to be determined are what effect each type and thickness of insulation will have on the cooling rate and how much it costs. The type and cost of various insulating materials can be determined using reference books and/or manufacturers' catalogs. With the technical data supplied by these sources and the theoretical equations you could find in a heat transfer text, you could calculate the expected effect of each type of insulation. If these results look promising, an experimental section of a typical bridge construction can be set up in a "cold-room" laboratory and actual tests run to verify the calculations. If more assurance is required, a test can be performed on an actual bridge.

Some of this work has actually been performed in research laboratories. What do you suppose the results showed? Decide on your answer before you read Section 22.

10

The results of both theoretical calculations and experimental studies showed that insulating the bottom of a bridge was <u>not</u> an effective way to prevent water freezing on its surface before the adjacent road froze. Without the heat supplied by the earth, the bridge surface still froze first most of the time. Thus, the chosen solution does not satisfy the objective of the project or the safety-economics criteria.

Choose a second solution to analyze from Section 20, then read Section 23.

The second best choice for a solution appears to be an electric heating system. Can you suggest the detailed analysis-synthesis-evaluation process that should be used for this choice? Determine your answer before you read Section 24.

This solution would appear to involve the use of a grid of wires imbedded in the surface of the bridge deck. Theoretical calculations can be used to design and predict the performance of this grid if we can estimate the expected air temperature and the amount of ice involved. Calculations could also be used to estimate the cost of a device to turn on the current whenever the temperature reached 32°F or, as an alternate, some man-controlled system for activation. Laboratory experiments could be used to verify the calculations, but an actual road test would probably be needed to check the method of imbedding the wires in the bridge surface to see that they would withstand the effect of traffic abuse.

All these tests have been performed. What do you think the results were? Decide for yourself before you read Section 25.

Calculations and tests performed in England showed that this was a practical, although expensive, way to eliminate the hazard of ice formation on a bridge surface.

Now that you have demonstrated that an electric heating system is an answer to the bridge freezing problem, you might think your job is through. It isn't. Before this system will be adopted, you must show that money spent on this device is better spent than on alternate improvements. For example, would the money spent on the bridge system prevent more deaths and injury than the same money spent on building new roads, repaving existing roads, building up shoulders, installing guard rails, railroad crossing lights or street lights? When the supply of

...oney is limited, as it always is, someone must decide how, when and where it should be spent. The engineer must be part of this decision-making process because he is the professional that can evaluate the alternates. Thus, your job on this project will be complete when you have gathered the data required to present the facts for the adoption or rejection of the new system. Until that is done, drivers will only have the sign that prompted this problem, "Bridge Freezes Before Road Surface," the spiked tires or chains they buy, and/or salt applied by the highway department as solutions to this problem.

Through this problem we have tried to indicate in a very brief way that many steps are required in the solution of an open-ended problem. In the more involved problems which follow you will be asked to perform these steps, learning as you go how to perform them with greater and greater skill. As you work on each problem, you will be expected to use what you already know and to learn new fundamental principles which are required to produce the desired result. In this sense, you will be operating exactly like a problem solver in the real world, who commands a broad background but must often search out new information to produce the best solution.

Before we move on to the next problem, it would be well for you to review the steps we have performed here. Although the order in which these steps occur varies from problem to problem, the steps themselves are an important key to successful work. Try to list the steps we have used, then check your list against our "Steps in Decision-Making."

Gather Information

This step is listed first because it may occur with <u>each</u> of the other steps in the decision-making process. It involves a search for pertinent information from the persons own background, books, printed matter, media, other people, experts and experimental work.

A. <u>Recognize a Problem</u>

Someone may ask you to solve a problem or you may discover it by yourself. In either case you must learn to look beyond the symptoms of the problem to find out what is wrong.

B. <u>State the Basic Objective or Goal</u>

The Basic Objective focuses your thoughts on the real problem to be solved. It should be a statement which is broad enough so no reasonable possible solution is eliminated.

C. <u>State the Constraints, Assumptions and Facts</u>

Constraints are factors which affect the outcome of the project and cannot be changed. Assumptions are applied to factors which can be changed to simplify the problem and make it solvable. Facts are statements of things that are known.

D. <u>Generate Possible Solutions</u>

This is the time for creative thinking. Don't prejudge ideas as they are generated; get all the different thoughts you can.

E. <u>Evaluate and Make a Decision</u>

Determine which possible solution is most likely to solve the problem.

F. <u>Analysis</u>

Separate the chosen possible solution into meaningful elements. Determine and gather the information you need to develop each element.

G. <u>Synthesis</u>

Combine the elements to create a detailed solution.

H. <u>Evaluate the Solution</u>

Does it satisfy the basic objective; is it feasible, practical, economical, safe, legal and moral?

I. <u>Report the Results and Make Recommendations</u>

Prepare a report that describes what you have done and decided. Think about the information needs of the man who will read the report. Don't write what isn't necessary.

J. <u>Implement the Decision</u>

What should be done now? Should we select a new basic objective, perform additional decision-making or research work, or begin to implement the recommended solution?

PROJECT I

BETTER NATIVE HOUSING

This project should take about four weeks to complete. It is
coordinated with Programmed Instruction 1 on the Energy Balance.

Introduction

John Newton, an engineer Peace Corps volunteer, and five other Peace Corps members have just settled into their seats on a jet plane for a nonstop flight overseas. In six hours, their plane will land in Port City. There they will be met by Harold Donnelly, the man who wrote the enclosed letter. This letter describes John's Peace Corps assignment. Although John is going to do the field work on this project alone, the five men on the plane will help him perform the design work for the project. As the plane drones on, the group begins to list the questions they will ask to get the information they need to begin their design work.

During the next few weeks, each of you in this class will play the role of John Newton, the project leader. The students who work with you will act as the other members of the Peace Corps team who help John plan. To begin your work form a team of four to seven students. Then each member of your group should read the letter to John Newton. Good luck, we hope your project is a success.

The Decision Maker

PEACE CORPS

TO: John Newton

SUBJECT: Overseas Assignment

FROM: Harold Donnelly, Peace Corps Supervisor

Because you are an engineer, you have been assigned to help the people in a small, primitive, isolated village deep in the rain forest develop their housing.

The village is located on the south shore of a very large lake, about 15 feet above the water's surface. To the west, the lake empties over a 200 foot waterfall to a river below. Fifty yards south of the village is a jungle-forest which ends 200 yards later in a steep cliff. The edge of the cliff is 20 feet below the village. The river is 100 yards from the base of the cliff.

The people of the village farm the area to the east of their straw huts. They cook their food over open fires in clay pots, which they themselves make, using materials found at the base of the cliff. The clay is carried to the village in baskets on a steep path up the side of the cliff.

Several years earlier, a missionary helped the people of this village build a permanent, brick meeting house. The people made the bricks themselves, using clay from the base of the cliff. The people are very pleased with this permanent building and would like their dwellings made of the same material.

The 13-day trip to the village is by Jeep followed by dugout canoe. A man from the village will be your guide. Your plans must be completed and approved in three weeks so you can begin your journey at the end of the rainy season.

A small library of books and facilities for any experimental work you may decide to do are available in the local university. In fact, during your stay here in Port City, you will be housed in a university dormitory. Although you will do all the field work for this project alone, the men who travelled to Port City with you will be available to help you plan the project.

Please submit your plans for the development of native housing to me as soon as possible. If I can be of help, do not hesitate to call on me.

Instruction 1.1--The Problem

After each man in your group has read the letter to John Newton, you should begin to discuss the project. In particular, we suggest you focus your discussion on a list of the questions you want answered. What additional information will you need to develop your plans? Your list of questions should be written, so your instructor can check it if he desires.

Project Leader.

When your list is nearly complete, select one man in your group to request Feedback 1.1 from your instructor. This man will then act as the Project Leader for this Instruction-Feedback set. (Note: this job should rotate through the group with each set.) The instructor may ask the project leader to describe the work the group has done. If they have done an adequate job of working toward an answer, the instructor will give the project leader the Feedback to read; then he returns to his group.

When the project leader rejoins his group, he has two very important jobs to do. First, his group may have had difficulty understanding the Instruction we have given. This is not unusual, instructions are often imperfect and different people interpret the same words different ways. Since the project leader has the Feedback, he should be able to clarify the job the group is supposed to do. Second, the project leader can provide some individualized help while the group is "thinking" about their work. In this case the project leader can help steer the group in the general direction we are going and make sure they do not miss any key points. This does not mean that the project leader should perform his job by simply giving the group our Feedback. Instead, he should ask questions which will lead the group in an appropriate direction.

The point of this project is to give you an experience in decision-making. The Feedback is provided to help guide your thinking. Don't misuse it.

Feedback 1.1

You have been selected to be the "project leader" for your group for this phase of the design project. Your first job as leader is to read the list of answers given below. This list will provide the information you need to answer your group's questions. Ask one member of your group to read a question from the group list. Then you can supply the answer to that question from what is given below.

If your group did not list one or more of these categories, you should try to ask them a question which will lead their thinking toward the missing topic or topics. If you feel some vital information is missing, check with your instructor.

When your group has discussed all these topics, give each man a copy of this Feedback and Instruction 1.2 to read. Proceed with Instruction 1.2 as soon as the group is ready.

Answers to Questions about The Project

1. Weather

The temperature ranges from 70° to 90° F. Two crops of vegetables and grains are grown each year. No crops are grown during the three-month long rainy season. During this season, floods often occur in the river valley.

2. Labor

The only labor available is from the villagers themselves. All are available except for normal daily duties associated with food preparation and retrieval. However, none of the villagers are available for a total of four weeks each year when they plant and harvest their crops. The village consists of one hundred families.

3. Transportation

Normally, all materials must be brought in by dugout canoe. Only in extreme emergencies has it been possible to arrange for air cargo drops, and then only after a messenger has travelled to Port City.

4. Natural Resources

There are unlimited amounts of fresh water available from the lake or river. The forest affords all types of trees and vines; however, because of the rainy season the wood rots quickly, so lumber is used only for short term projects. The area is rocky, except for sections of sandy soil where the villagers farm, and clay at the base of the cliff.

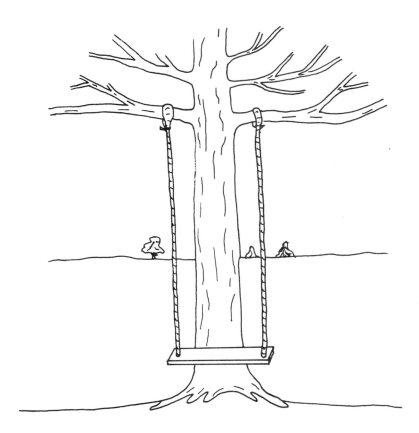

The Setting

5. Timetable

John Newton will be at the village for only <u>fourteen months</u>. The project should be completed before he leaves.

6. Tools

Common farming hand tools are available: axes, hammers, saws, hoes, shovels, forks, rakes and flails.

7. Budget

Only one hundred dollars is available. It can be spent in whatever way desired.

8. Other Factors

(1) The present dwellings are damp in the rainy season.

(2) Straw huts breed bugs.

(3) Straw huts disintegrate in about three years.

(4) Aluminum roof sheeting will be made available by the government if the villagers complete the brick portions of the new houses.

Instruction 1.2--Basic Objective

Now that your group has gathered the available information about the village and the project, you should be ready to state the basic objective of your work.

<u>The basic objective should clearly define the desired result of the project and nothing else.</u> One sentence should be sufficient. It should not include any statement of how the project will be performed.

- -

Your group should now discuss and define the basic objective. After you have discussed what you think is the basic objective of your assignment, write it down for future reference. In actual engineering practice an engineer would use a "Log Book" for this record. We have provided Log sheets on the two pages which follow. Put your objective on this log, then have a <u>new</u> project leader read Feedback 1.2.

PROJECT LOG BOOK

1. Our Basic Objective or Goal is

2. The Constraints, Assumptions and Facts are

3. The Possible Solutions we Generated are

4. We Evaluated and Made these Decisions

5. <u>Our Analysis is as follows</u>

6. <u>The Result of our Synthesis is</u>

7. <u>Our Evaluation of this design shows</u>

8. <u>We Recommend</u>

You have been selected to be the project leader of your group for this phase of the design project. As part of your task, you should first study the basic objective stated below.

Basic Objective

To design and build permanent brick dwellings for the people of the village.

Consider your group's basic objective; is it as simple as ours? If the group's objective is more detailed than the one stated above, it may, in fact, restrict the choice of possible solutions to the problem.

Note carefully that your basic objective does not have to agree exactly with the one given here. In fact, you may find that the basic objective given above is not as good as the one your group has developed. If you believe this is the case, don't hesitate to challenge this objective. There is nothing magic about words that have been printed. Both authors and instructors make mistakes.

Instruction 1.3--Constraints, Assumptions and Facts

When you solve a complex, open-ended problem you will generally discover that there are a few factors which are very important to the problem solution. Most of hese factors can be identified as constraints, assumptions and facts. We suggest that you prepare a list of these factors so they are available for easy reference as you proceed with your work.

The constraints on your list will usually describe limits or extremes which must be considered in the design. For example, one limit in this project is the $100 budget.

The facts on your list should come from the information you have available. For example, you know that the rainy season is three months long. This may or may not prove to be a constraint.

If some of the information you need is not readily available, you may have to make an assumption to proceed. For example, you do not know the size of a typical native family; therefore, you may have to assume that each group includes four or five or six people.

At this point your group should prepare a written list of constraints, assumptions and facts. Then, a new project leader should check the Feedback.

The Problem

Feedback 1.3

The project leader should check our list of constraints, assumptions and facts against that generated by his group.

1. The only labor available is that of the villagers themselves. (C)
2. They can work only eight months of the year. (C)
3. Only $100 is allotted for the project. (C)
4. Available transportation is low in capacity and slow. (F)
5. Only hand tools are available, no machinery. (F)
6. There is a three-month rainy season. (F)
7. The job should be completed in fourteen months. (C)

Did your group prepare a comparable list of factors? If any were missed, lead your group through a discussion until they arrive at them. If your group had other factors listed, consider carefully whether they result from these or are so obvious that they need not be stated. For example, bricks may not be flown in, is a constraint which is too obvious to be stated. Certainly the $100 limit eliminates this possibility. Bricks cannot be made from concrete is also an obvious constraint for the same reason.

You may have noticed that our list of factors does not include any assumptions. You may also have had trouble generating these factors at this point in the project. In fact, we expect you will discover that assumptions cannot be generated until you have a need for them, which will occur later in the project when you are ready for detailed calculations. For example, when you do your calculations you may need to know how much clay a native can dig each hour. Unless you can find this information in a reference book (or know it from personal experience), you will have to make an assumption.

When the group has finished discussing the constraints, each man should read the Instruction for the next step.

Special Instruction--Programmed Instruction

The constraints you have listed describe limiting factors which may prevent you from achieving your objective. Therefore, you should check each constraint carefully. In this project, the key constraints appear to be related to the "work" the natives can do in the time available. Thus, constraints 1, 2 and 7 on our list should be checked carefully.

1. The villagers will do the work.
2. They can work only eight months/year.
7. The job should be completed in fourteen months.

To prepare yourself to decide whether the natives can perform this work you will

23

need to know something about the work a man can do. To help you get the required background, we have prepared some programmed instruction on the subject of Conservation of Energy. We suggest that you begin to study this material now, so you will be prepared when you need this information.

You should begin your work with the first program included with this design project, a program on "Shaft Work and Force." The concepts developed in each program will be introduced by a discovery-experiment which relates each new concept to an appropriate physical activity. The programs themselves should be studied at home. If your project supervisor has not already given you the programmed material for this project, ask him for your set.

Although some class time may be required to perform experiments or discuss the homework problems given with the programs, you should plan to continue with the design process in class. This combination of study and design is the same pattern used by all professionals who face a problem which requires them to review old concepts or learn new ones to complete their design work.

Now move on to the next Instruction.

Instruction 1.4--Generate Possible Solutions

If you check your list of decision-making steps you will see that the next step is to generate possible solutions which will achieve the basic objective: to build permanent brick dwellings for the natives. Although this is a logical next-step, we would like to suggest that you perform another task which will make the generating process much easier. The task is to analyze or breakdown your basic objective into components, or what might best be called sub-objectives. Each sub-objective will describe a component of the problem to which you can generate possible solutions. The sub-objectives you need here are related to "what" and "where." See if your group can develop these two sub-objectives. When you have identified these two components the project leader should check the Feedback.

One Possible Solution

What--The fact that you plan to build permanent brick dwellings is not a full statement of the problem. One of the important decisions that must be made is <u>what</u> sort of dwelling to build.

Where--A second important decision that must be made is <u>where</u> to build the dwellings.

If your group has not arrived at these decisions, see if you can lead them to identify these two sub-objectives.

Note: Although we expect your group to follow the general design pattern given for this project, there is no reason why the group must agree exactly with each step. For example, at this point you might choose to consider building materials other than the clay. We would encourage you to develop your own ideas and to explore other possibilities. The material that follows develops one possible solution and shows the pattern for this type of design work. Most of this can be used as a guide for your work even if you follow a slightly different design path.

Instruction 1.5--Generate Possible Solutions--Where

Each of the sub-objectives you have developed can be treated as if it were a basic objective for the project. This means all the decision-making steps can be applied to each sub-objective. For example, if you start your work with "where" to build the dwellings, you can specify the constraints, assumptions and facts that apply, generate possible solutions, evaluate the solutions, etc. Since your group has already specified a set of constraints, we suggest you start your work with the process of generating possible solutions for "where" to build the village.

Your group should now <u>dream-up</u> as many possible solutions as they can for this sub-objective. <u>List</u> all the ideas that are developed; do not reject any suggestion. Some idea which appears unrealistic at first may prove to be the best idea later, or it may stimulate a better idea from someone else.

When your list is complete, a project leader should read the Feedback.

Feedback 1.5

We have identified three possible solutions. Your group may have identified
more and better solutions and we hope they have. However, if your group has somehow
missed one of our solutions, we hope you will attempt to lead them to add this to
their list. Our solutions are:

1. Move the village to the river level near the clay.
2. Build the village where it is now.
3. Do not build new dwellings at all.

As soon as your group has finished their list of possible solutions, give each
member this Feedback and the next Instruction to read.

Instruction 1.6--Evaluate the Possible Solutions

The next step in the decision-making process is to evaluate your possible solutions and rank them in order from the best to the poorest. To help you perform this
step we have provided a set of Design Decision Tables.* Each member of your group
should receive:

1. one copy of the Instructions for the Design Decision Table.
2. three copies of the Design Decision Table form.

Ask the group to consider and discuss the factors listed in the Table. Then
ask them to use the Table to analyze each of their possible solutions, factor by
factor, either acceptable or unacceptable. Be sure the group records a reason for
their rating in the "Explanation Column."

When your group is almost finished with its work, a new project leader should
read Feedback 1.6.

---

*Note: Design Decision Tables are used to help you choose between possible solutions.
Sometimes a solution can be eliminated without using the table; for example, if you
know a solution does not satisfy the basic objective, then you do not need to fill
out a form for it. In addition, you should realize that the tables are not required
for all decisions. Sometimes the alternates are such that you can choose the best
one without the use of any tables. When you write your report, you will find the
tables a convenient method of presenting your decisions. However, even if the
tables are included, your decisions should be summarized in the written portion of
your report. If you do not use the tables, then you should describe each decision
in the report.

26

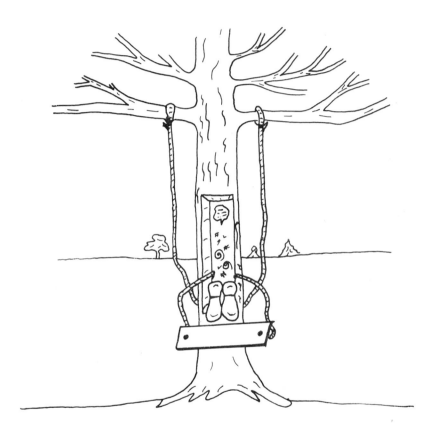

Test Results

DESIGN DECISION TABLE

Objective _____

Possible Solution _____

Factor	Acceptable	Unacceptable	Explanation
1. Does the solution satisfy the basic objective of the project?			
2. Is the solution theoretically feasible?			
3. Is the solution practical?			
4. Does the solution satisfy all the social and ecological factors involved?			
5. Is the effort (time, energy, money) invested in the project worth the value received?			
6. Can the project be completed in a reasonable time?			
7. Is the solution legal?			
8. Are there any special side benefits produced by the solution?			

27

DESIGN DECISION TABLE

Objective _____

Possible Solution _____

Explanation

Factor	Acceptable	Inacceptable	
1. Does the solution satisfy the basic objective of the project?			
2. Is the solution theoretically feasible?			
3. Is the solution practical?			
4. Does the solution satisfy all the social and ecological factors involved?			
5. Is the effort (time, energy, money) invested in the project worth the value received?			
6. Can the project be completed in a reasonable time?			
7. Is the solution legal?			
8. Are there any special side benefits produced by the solution?			

28

DESIGN DECISION TABLE

Objective _____

Possible Solution _____

Factor	Acceptable	Unacceptable	Explanation
1. Does the solution satisfy the basic objective of the project?			
2. Is the solution theoretically feasible?			
3. Is the solution practical?			
4. Does the solution satisfy all the social and ecological factors involved?			
5. Is the effort (time, energy, money) invested in the project worth the value received?			
6. Can the project be completed in a reasonable time?			
7. Is the solution legal?			
8. Are there any special side benefits produced by the solution?			

INSTRUCTIONS FOR THE DESIGN DECISION TABLE

1. Does the solution satisfy the basic objective (and/or sub-objectives) of the project?

    Answering this question provides you with an opportunity to check the solution against the basic objective. You may have drifted away from the basic objective without realizing it.

2. Is the solution theoretically feasible?

    Carefully check your proposed solution to see if it violates any known scientific principle.

3. Is the solution practical?

    It may be scientifically possible to create a system to move enough clay in one single, monstrous piece. However, it is not practical to build such a system.

4. Does the solution satisfy all the social and ecological factors involved?

    You must consider how the possible solution affects the human beings involved. Factors to be studied here include safety, pollution, social customs, and psychology. It may be feasible to build small apartments to house 30 people, but it is unlikely that the villagers would remain happy living there.

5. Is the effort (time, energy, money) invested in the project worth the value received?

    For example, is it worthwhile to build an outdoor unheated swimming pool in Alaska, where it might be used only one month a year?

6. Can the project be completed in a reasonable length of time?

    For example, it would be unreasonable to take three days to pitch a tent, five years to build a house, or six weeks to repair a leak in a water pipe.

7. Is the possible solution legal?

    Now, after questioning all other aspects of the solution, one can consider its legality. The legal questions concern licenses, permits, patents, government regulations and standards. The laws governing these areas can sometimes be changed when a strong case is presented.

8. Are there any special side benefits produced by the solution?

    This is a bonus factor, not necessary in the design decision, but valuable in comparing equally acceptable solutions.

<u>Feedback 1.6</u>

As your group works, the Project Leader should compare their answers with those on the Design Decision Tables we have given on the next three pages. Use our answers to guide the group toward a complete rating of all the possible solutions. If one solution includes a factor which is rated unacceptable, ask the group how the design could be changed, or what new design could be used to bring that rating up to an acceptable level.

When the group completes checking their tables, give them a copy of this Feedback and our Tables to read. After any further discussion is completed, each member of the group should read the next Instruction.

DESIGN DECISION TABLE

Objective   Where to Build

Possible Solution   Move the Village to the River Level

Factor	Acceptable	Unacceptable	Explanation
1. Does the solution satisfy the basic objective of the project?	X		This solution does satisfy the basic objective.
2. Is the solution theoretically feasible?	X		There is no apparent violation of scientific principle.
3. Is the solution practical?	X		Moving the village site is fairly simple since the houses are to be rebuilt in any case.
4. Does the solution satisfy all the social and ecological factors involved?		X	This solution neglects the initial reason for building the village on top of the cliff, where it is near the farmland and the lake, and not likely to flood. The villagers would also have to abandon their meeting house.
5. Is the effort (time, energy, money) invested in the project worth the value received?		X	The villagers would have a new village, but would now have to travel up the cliff to work in the fields.
6. Can the project be completed in a reasonable time?	X		The clay is immediately available in large amounts. The natives do not have to carry it up the cliff.
7. Is the solution legal?	X		No legal problems are likely to appear this far away from civilization.
8. Are there any special side benefits produced by the solution?			There are no special side benefits.

Objective __Where to Build__

Possible Solution __Where the Village is Now__

DESIGN DECISION TABLE

Factor	Acceptable	Unacceptable	Explanation
1. Does the solution satisfy the basic objective of the project?	X		This solution does satisfy the basic objective.
2. Is the solution theoretically feasible?	X		There is no apparent violation of scientific principle.
3. Is the solution practical?	X		Yes, but there is a great deal of clay to move up the cliff path.
4. Does the solution satisfy all the social and ecological factors involved?	X		Yes, this location is apparently where the natives prefer to live. But this fact should be checked.
5. Is the effort (time, energy, money) invested in the project worth the value received?	X		Yes, if the villagers will be better off in these new dwellings. This should be checked.
6. Can the project be completed in a reasonable time?	X		It will take a considerable amount of time and effort to transport the clay.
7. Is the solution legal?	X		The villagers own the path and are not affected by any regulations.
8. Are there any special side benefits produced by the solution?			There are no special side benefits.

Objective __Where to Build__

Possible Solution __Do Not Build New Dwellings__

DESIGN DECISION TABLE

Factor	Acceptable	Unacceptable	Explanation
1. Does the solution satisfy the basic objective of the project?		X	No, but we can't be sure that the natives really want new dwellings; i.e., will they live in them after they are finished?
2. Is the solution theoretically feasible?	X		No scientific principles have been violated.
3. Is the solution practical?	X		Yes, it saves a lot of effort. Perhaps there is something better we can do for the villagers in the time available.
4. Does the solution satisfy all the social and ecological factors involved?	X		Yes, the natives may be safer and happier in their straw dwellings. We don't know what impact these new dwellings may have on the people.
5. Is the effort (time, energy, money) invested in the project worth the value received?	X		Yes, if the people are better off this way.
6. Can the project be completed in a reasonable time?	X		Yes.
7. Is the solution legal?	X		Yes, but you may have some explaining to do to your Peace Corps supervisor.
8. Are there any special side benefits produced by the solution?			Yes, we can do something else for the natives, something they actually want.

Instruction 1.7--Rank the Possible Solutions

      Now your group should rank their possible solutions and determine which one is most likely to result in the best solution to the problem.

      When the group has completed this ranking process, give each member the next Feedback to read.

Feedback 1.7

Shown below is a ranking based on the three possible solutions we analyzed.

    First: Build the village where it is now.

    Second: Move the village to the river level near the clay.

    Third: Do not build new dwellings.

Although this order appears to be the best at this point, you should recognize the need to find out what the natives want before you proceed. This indicates that a gathering information step is required before you actually proceed with the construction work. Thus, when you get to the village, you may discover that the third solution is, in fact, the proper one. However, since you have time to plan now, you should proceed with your best solution.

Instruction 1.8--Generate Possible Solutions--What

Now that you have decided where to build the dwellings, your group should generate possible solutions for the problem of what type and size of dwelling to build. Your own experience should be of some help here. We will not use Decision Tables at this point, a simple list will do. What kind of brick dwellings might you build for the natives?

When your list is completed, have a new project leader read the Feedback.

A New Possible Solution

Feedback 1.8

The dwellings you might build are:

I. Rooms
   A. 1 room
   B. 2 rooms
   C. Several rooms

II. Style
   A. 1 family
   B. 2 family
   C. Apartment house
   D. Row houses
   E. Square or rectangular

III. Floors
   A. 1 floor
   B. 2 floors

IV. Roof
   A. Flat roof
   B. Sloping roof
   C. Peaked roof

Instruction 1.9--Gathering Information

The many options available to you should show that this part of the problem is not as simple as the last part. For example, a major factor to consider when you make this decision is the constraint that the natives must carry the clay up the cliff path and build each house. In addition, you cannot decide what kind of dwelling to build without knowing something about how people (or the natives) use their dwellings. In general, what do people do in their dwellings? Prepare a written list of the factors which should be important. Once you have generated ten or more factors (we have twenty-five) proceed as before by having one member check the Feedback.

The following information on "needs" related to a home were taken from the final report of a NASA project called "Starsite," Toward a Decision Making Mechanism for Housing.*

The functions of a house are to:

1. Provide a place to sleep.
2. Provide shelter from weather: wind, rain, heat, cold, sun.
3. Provide protection from the criminal element.
4. Provide storage.
5. Provide for disposal of personal effluent: excrements, CO_2, body heat.
6. Provide personal hygiene facilities.
7. Provide for cooking usage.
8. Provide area for relaxation and/or personal development.
9. Provide place to eat.
10. Provide comfortable atmosphere: temperature, humidity.
11. Allow for sexual activity.
12. Provide isolated space (reading, thinking, etc.).
13. Insulation from noise.
14. Protection from pests.
15. Provide lighting for all activities.
16. Provide space and surfaces for writing and working.
17. Provide space for family interaction.
18. Provide ability to use various small appliances.
19. Communication (visual) with immediate exterior.
20. Communication with remote exterior (telephone, radio, and TV).
21. Safety (from potential hazards within and without the home).
22. Maintain adequate pressure and mixture of atmosphere.
23. Allow for decoration.
24. Allow for good access to outside.
25. Provide for convenient, sheltered parking.

Your group must decide which of these elements are appropriate for the native dwellings you plan to build. Then move on to the next Instruction.

_____

*NASA, 1971

Instruction 1.10--Analysis

At this point your group will have completed the evaluation of the possible solutions for each of the sub-objectives so you now know what type of dwellings you plan to build and where you plan to build them. This means it is time to move on to the next step in the decision-making process, which is an analysis of the chosen solution. In the analysis phase you should determine all the significant elements or variables which must be considered. This may overlap partly with the synthesis phase which follows, where you organize all the elements into a final solution. And these two phases often occur simultaneously with an evaluation phase, so your solution may be revised while you work.

Your group should now begin the analysis phase of their design work by identifying the basic elements involved in building new brick dwellings on the site of the present village. We are not concerned here with a detailed list of elements, but rather a set of broad categories. What general factors must be considered in the process of obtaining materials and building the type of brick house you have chosen?

We see four basic elements in this systems design. The natives must:
1. Dig the clay.
2. Carry the clay or bricks up the cliff path and move it to the village site.
3. Make the bricks.
4. Build the dwellings.

Instruction 1.11--Detailed Analysis

Your group should now proceed with a detailed analysis of this systems design and determine all the factors that are important to each basic element. This is an extremely important step so don't rush it, take plenty of time and do it right. When the group has completed their analysis, have your next project leader read the Feedback.

Your group should have completed a detailed analysis such as that shown below. If any items are missing from the group's list, ask questions which will lead them to complete their list.

Clay-Housing System--Basic Elements and Detailed Components.

1. Digging the clay
   a. Quantity dug per hour or day
   b. Number of men required
   c. Tools required
2. Carry the clay up the cliff path and move it to the village site.
   a. Weight and volume per load
   b. Number of men required
   c. Equipment required
3. Making the bricks
   a. Make the bricks before moving the clay up the cliff.
   b. Make the bricks near the village.
   c. Time required
   d. Is clay the only material required?
   e. Is mortar required?
4. Build the dwellings
   a. Amount of clay required
   b. Number of men required
   c. Time required
   d. Tools required

Instruction 1.12--Information Sources

The detailed analysis step should alert your group to a need for information about the process they must design. For example, the design has actually proceeded quite far without our having checked the quantity of clay which must be carried up the cliff path. This factor should have been determined quite early because it may have a significant effect on the design. For example, if only 10 pounds of clay are needed for each house, each man can easily carry his own load up the path. Of course, you know that quite a bit of clay is required, even if you haven't calculated a number. But, before you can proceed with the design you must have this and other detailed information.

In general, there are four ways in which you can obtain the information you need. We would like you to identify these four sources and have provided the questions given below to help you do it.

1. Why was John Newton chosen for this project?

2. How can you determine the exact definition of a word such as ecological?

3. How can you find your way to the village where you are assigned?

4. How can you determine the weight per unit volume of the clay?

Although each of these questions should suggest a different information source, there may be some overlap.

When your group has identified these four sources, the Project Leader should read the Feedback.

eedback 1.12

The four basic sources of information which your group should have identified
are:

1. Your own previous education, training or experience. (John Newton was
   chosen for this project because he has a background as an engineer.)

2. Books, other printed matter or media. (The dictionary can give you the
   exact answer to this question.)

3. Other people--experts. (The expert who can show you the way to the village
   is the native guide.)

4. Experimental work. (One way to determine the density of the clay is to
   measure it.)

Instruction 1.13--Analysis-Synthesis-Evaluation

Your experience should include some of the background required for the work you
now must do. For example, in your earlier education you should have learned how to
calculate the volume of clay required for the 100 houses you will design and build.
In addition, the energy balance concepts you have learned as part of this project
should help you perform the "work" calculations involved. Some of the other information you need will probably come from the library.

Since you may not have an expert available to you, we have obtained some of the
additional information you will need. For example, we have determined that the
caloric intake of a native is only about 80% of our intake of 2520 kcal per day.
The experts also say that only 5.3% of this energy is available for the work each
native may want to do. The rest is used to maintain body functions or wasted.

Your job now is to complete the project. The remaining steps involve detailed
calculations, the synthesis of a system, and the evaluation of your solution. As
you work you should be alert for better ways to do the job. In addition, you should
be sure that you have satisfied the basic objective of your work and kept within all
the constraints.

Before you leave this material to complete your work, read the next
Instruction.

Time for Reevaluation

Your detailed design will be the Feedback to this Instruction.

Instruction 1.14

Oral Report.

As soon as the details of your design work are complete, each group should be prepared to present an oral report to their supervisor (your instructor and/or the whole class). This report should summarize all the major decisions made by the group. It should not take longer than ten minutes.

The oral report is a commonly used technique in the real world of the professional engineer. It has two main functions:

1. It requires the designer to bring together all the details which represent the total design.

2. It provides a check on the work--others can judge if the designer has developed a completely reasonable, workable, and acceptable solution to the original objective.

Written Report.

When your design work is completed, each member of the group should prepare an individual, written report on the project. <u>This report should describe the decisions made by the group</u> (as if each person were the secretary of the group). However, you may also point out ideas which you believe the group should have used. We would suggest that each report contain the following parts.

1. Cover Page
2. Introduction
3. Basic Objective and Sub-Objectives
4. Design Constraints
5. Possible Solutions
6. Preliminary Evaluation (including the Design Decision Tables)
7. The Final Choice
8. Analysis (calculations)
9. Synthesis (calculations and sketches of the final design)
10. Evaluation
11. Recommendations

In other words, we suggest that you use the decision-making steps as the basis for your report outline. We expect that you will find this a useful structure for your work.

You will receive the next Feedback when you turn in your completed report.

The Final Solution

The feedback to the job you have just completed is our copy of an acceptable Final Report, which is given on the next few pages.

Read the next Instruction before you read our report.

Instruction 1.15--Evaluation of Your Report

Your job now is to evaluate your Project Report by comparing it to ours. If your report does not measure up, rewrite it (but don't copy ours--use your own words). If your report is as good or better than ours, grade it. That's right, give yourself a grade. Your project supervisor will review your report and determine if the grade you gave yourself is reasonable. If your grade is too high, he may ask you to rewrite your report until it meets the grade you have assigned to your report.

We believe that the self-evaluation of your report is an important part of your education, because eventually you must learn not only to write well, but also to evaluate your own work. Engineers must communicate well because they have important information, designs and ideas to communicate.

FINAL REPORT

New Dwellings for a Native Village

by _____

Group _____

Class _____

Date _____

Project Supervisor _____

Introduction

This report describes my work on a Peace Corps assignment to develop better housing for a group of natives who live in the rain forest, 13-days travel from Port City. Before leaving for the village I was asked to prepare preliminary plans for the project. My work is based on the limited information available at this time; therefore, some of the decisions that have been made must be re-examined when I reach the village.

The site of the present native village is located between a large lake and a 200 foot cliff. The people live in straw huts. Their desire for new dwellings appears to have been stimulated by the brick meeting house a missionary helped them build several years earlier. The clay for these bricks was carried by the natives from the base of the cliff to the village. The job at hand is much more complex, because the natives want dwellings for 100 families. Although the government is unable to aid the villagers financially, they have provided an incentive; they will supply aluminum roofing sheets when the brick portion of the dwellings are completed.

My job was to plan for the development of these new dwellings.

Basic Objective

The basic objective of this project is to design a system which will provide better housing for the people in a native village.

Sub-Objectives

The basic objective was divided into three sub-objectives.

A. What material to use.

B. Where to build the dwellings.

C. What type of dwelling to build.

Each of these sub-objectives is examined in the material which follows.

Design Constraints

The basic objective is subject to the following constraints.

1. The only labor available is that of the villagers themselves.

2. They can only work for 10 months over the allotted period of time, due to the three month rainy season and the harvest period.

3. Only one hundred dollars is available for the project.

4. The job should be completed in fourteen months.

A. First Sub-Objective--What Type of Material To Use

A.1--Possible Solutions

a. Brick

b. Rock

c. Wood

d. A combination of all three materials

A.2--Evaluation

Rock and/or wood could be used to build the new dwellings. However, the rock might be very difficult to obtain and shape and mortar would probably be required to hold it in place. Although wood rots, it might be possible to find a preservative in the jungle.

A.3--Decision

Since the clay is available, the natives have had experience with this building material, and brick houses were requested, we finally decided to make brick dwellings. However, some parts of each dwelling, such as the lintels may be made of rock. Wood, if it can be preserved, may be used as part of the roof support.

B. Second Sub-Objective--Where to Build The Dwellings

B.1--Possible Solutions

a. Move the village to the river level near the clay.

b. Build the village where it is now and have the villagers carry the clay up the cliff.

c. Do not build new dwellings at all.

These possible solutions were evaluated on a set of Design Decision Tables given in the Appendix.

B.2--Evaluation

Moving the village to the river level was rejected for several reasons. First, the river may flood and drive the people from their homes. Second, if the village is moved the people will have to travel up the cliff to get to both their fields and the meeting house. This is bound to involve much greater effort than that required to move the clay up the cliff.

Not building new dwellings was considered as a real possibility which should be checked before construction begins. It is entirely possible that the natives will not want to live in brick dwellings.

B.3--Decision

Our decision was to build the native village where it is now. However, we expect to check the attitude of the natives before any work is done. In addition, it seems reasonable to check their attitudes by completing one "test house" before a large scale effort is launched.

C.   Third Sub-Objective--What Type of Dwelling

   C.1--Possible Solutions
      a.   Multiple-family dwellings
      b.   Single family dwellings
      These two possible solutions represent the extremes available to us. They
      are evaluated on a set of Design Decision Tables given in the Appendix.

   C.2--Evaluation
      Multiple-family housing requires fewer bricks and consequently less mate-
      rial must be moved and prepared. However, it seemed reasonable to assume
      that the natives might have great difficulty adjusting to the problems
      associated with this type of living arrangement. The natives now live in
      single family huts. It may be that they would be happier with some other
      arrangement, i.e., more than one room or communal living. However, we
      cannot determine these attitudes until we get to the village and have a
      chance to talk to them and observe their life style.

   C.3--Decision
      With the information presently available, we decided to build single
      family units.

Analysis-Synthesis of the Dwelling

   In the absence of information on the native's life style we had to make
arbitrary decisions about the size and shape of the house we plan to build. However,
these decisions allow us to make preliminary plans and even if changes must be made
later, what we have done can serve as an estimate of the work involved.

Size and Shape

   Our choice for a dwelling was a square, 16 feet x 16 feet, two-room house with
8-inch thick walls. The front wall is 9 feet high, but the back wall is only 8 feet
high to provide a sloping roof for water drainage. An interior dividing wall was
added to provide support for the roof. One window-opening will be left in each wall
to permit flow-through ventilation and to admit light. One door-opening will be
provided in the front wall. Stone from the cliff may be used for the support over
each opening or a brick arch might be used. The floor will not be paved. Bamboo
drain lines might be added at the base of each wall to provide an outlet for rain
water.
   The aluminum roof materials for these dwellings should be at least 18 feet
long. The exact details of the government's plans should be checked to be sure
they can deliver this material.

Note: A square configuration was chosen because square houses require fewer bricks than rectangular houses to give the same floor area. However, a round house requires still fewer bricks, about 11-1/2% less. Therefore, an igloo type building might also be considered. This construction has the additional advantage of eliminating the need for a roof.

Bricks

Even though we have specified an 8-inch thick wall, we have not yet specified the size of a brick. It may be that the natives will want to use a brick similar to those they used before. The key factors to consider in this decision are the stability of the walls and ways to make the clay go as far as possible. For stability, the bricks might be made 8-inches wide, with some laid across the wall and others along the wall. To extend the clay, the bricks might be made hollow. If these hollow spaces line up, additional clay might be used to lock the bricks together. We might also consider a ridge and groove arrangement to provide locking and to prevent water from running between the bricks.

Making the Bricks

If the bricks are baked (rather than sun dried) they may be made either on top or at the bottom of the cliff. This decision will depend on the previous experience of the natives, the loss in weight that occurs during baking, and the availablity of fuel.

Detailed Analysis of the Clay Required

1. The amount of clay required

   One Wall

   The average wall is 16 feet long, 8.5 feet high and 0.667 feet thick.

   $$\frac{\text{Volume of clay}}{\text{wall}} = (16 \text{ ft})(8.5 \text{ ft})(0.667 \text{ ft}) = \underline{90.7 \text{ cu ft}}$$

   Four Exterior Walls + One Partition Wall

   $$\frac{\text{Total volume of clay}}{\text{house}} = \frac{(90.7 \text{ cu ft})(5 \text{ walls})}{\text{wall}} = \underline{453 \text{ cu ft}}$$

   100 Houses

   Total volume of clay = 453 cu ft x 100 = $\underline{45,300 \text{ cu ft}}$

   Assume the clay weighs $\frac{120 \text{ lb}}{\text{cu ft}}$

   Total weight of clay = 45,300 cu ft x $\frac{120 \text{ lb}}{\text{cu ft}}$ = $\underline{5,436,000 \text{ lbs clay}}$

   This figure does not take into account the openings in the walls. However, since extra material may be needed to build foundations for the

walls and/or to allow for material losses, this error will be neglected.

2. The Working Time Available

The natives are available for work eight months out of a twelve month period. This is due to a three month rainy period and a one month period required to plant and harvest crops. Since the rainy season is just ending, the natives will be available for a maximum of ten months of work. We will assume that the natives can be expected to work four weeks each month, five days per week and eight hours per day. Therefore, the total time available for this work is

$$10 \text{ months} \times \frac{4 \text{ weeks}}{\text{month}} \times \frac{5 \text{ days}}{\text{week}} \times \frac{8 \text{ hours}}{\text{day}} = \underline{1600 \text{ hours}}$$

3. Amount of Clay Moved Per Hour

$$\frac{5,436,000 \text{ lbs clay}}{1600 \text{ hours}} = 3397 \frac{\text{lbs clay}}{\text{hr}} \cong 3400 \frac{\text{lbs clay}}{\text{hr}}$$

4. Amount of Work a Native Can Do

Our caloric intake = 2520 kcal/day = $\underline{7.78 \times 10^6 \text{ ft-lb}_f/\text{day}}$

Native intake--80%

$$(0.8)(7.78 \times 10^6 \text{ ft-lb}_f) = \underline{6.224 \times 10^6 \text{ ft-lb}_f/\text{day}}$$

Useful energy available--5.3%

$$(6.224 \times 10^6 \text{ ft-lb}_f)(0.053) = 3.298 \times 10^5 = \underline{329,800 \text{ ft-lb}_f/\text{day}}$$

5. Clay Carried by One Native

We can assume that a typical native weighs 150 lb_f. We can also assume that he can carry 50 lb_f of clay each trip. If this is the case, the minimum energy required per trip is

$$(200 \text{ ft cliff})(50 \text{ lb}_f \text{ clay} + 150 \text{ lb}_f \text{ man}) = 40,000 \text{ ft-lb}_f$$

Note: The inefficiency of this process should be obvious. Three fourths of the energy involved in this lifting process is wasted as the man lifts himself up the cliff. Each native could move much more clay if he did not have to lift himself up with it. In fact, he could move up to 150 lb_f more clay with the same energy output.

The number of trips the native can make is a function of the total energy he has available.

$$\frac{329,800 \text{ ft-lb}_f \frac{\text{available}}{\text{day}}}{\frac{40,000 \text{ ft-lb}_f}{\text{trip}}} = \underline{8.245 \text{ trips per day}} \text{ or } 8 \text{ trips per day}$$

If the native carries 50 lb_f of clay each trip and he makes one round trip each hour, then it will take

$$\frac{3400 \text{ lbs clay}}{\text{hr}} \times \frac{1 \text{ man}}{\frac{50 \text{ lbs clay}}{\text{hr}}} = \underline{68 \text{ men}}$$

sixty eight men to carry the clay up the cliff path.

6. Digging the Clay

   We will assume that a man can dig 500 lbs of clay per hour. In this case it will take

   $$\frac{3400 \text{ lbs clay}}{hr} \times \frac{1 \text{ man}}{\frac{500 \text{ lbs clay}}{hr}} = \underline{6.8 \text{ men}}$$

   about seven men to do the digging.

7. Make the Bricks and Build the Houses

   Since we have now accounted for 68 + 7 = 75 men, there are 25 men left to make the bricks and build the houses. This may be enough men to do the job.

Final Decision

A preliminary design has been completed which involves the building of square, single-unit, clay brick houses at the present village site. The clay for making the bricks will be carried up the cliff path.

Recommendation

This study indicates that the majority of the workers will be involved in carrying the clay up the cliff. Thus it is recommended that a more efficient clay transportation system be designed.

Appendix

Because a completed set of Design Decision Tables was given in the Guided Design project, they have not been duplicated here.

PROJECT II

BUILDING ACCESSIBILITY

This project should take about four weeks to complete. It is coordinated with Programmed Instruction 2 on Systems With Mechanical Advantage.

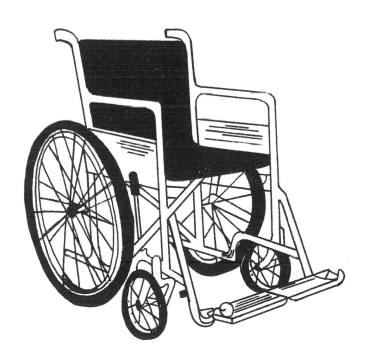

Introduction

Although <u>Tom Blaskovics won't graduate from engineering school</u> for another year, he has been hired for the summer by the State Department of Vocational Rehabilitation. His first assignment is to provide assistance to one of the <u>rehabilitation counselors, Chuck Stuart</u>.

For the past six months Mr. Stuart has been working with a man who had a back injury and is now permanently assigned to a wheelchair. The <u>injured man, Mr. Robert Masson</u>, has been re-trained as a bookkeeper and a local employer has agreed to hire him, but the small factory where he would work is located in an old building which is <u>not</u> accessible to a wheelchair--it has steps leading to all the entrances.

Instruction 2.1

Tom meets with Mr. Stuart in his office to discuss the situation. During this conversation, what would Tom suggest should be done about this accessibility problem.

If Tom forgot that the proper way to begin the solution of an open-ended problem is to define the problem and state the basic objective, he and Mr. Stuart might have discussed a set of possible solutions for this accessibility problem. For example, they might have decided to:

1. Move the business to a modern building where there are no steps.
2. Cover a set of steps with plywood to form a ramp.
3. Get several men to carry Mr. Masson and his wheelchair in and out.

Instruction 2.2--Basic Objective

It turned out that Tom did remember that the proper first step is to define the problem and state the basic objective. If you were in Tom's shoes, what objective would you have specified?

When your group has completed their objective (or if you have already developed one), a project leader should check the Feedback.

Feedback 2.2--Basic Objective

Your statement of the objective for this project may take a variety of forms. For example, your statement might be similar to one of the three given below.

1. To make the building safely accessible to all people who cannot climb stairs.

2. To get Mr. Masson to and from his new job.

3. To modify at least one set of stairs for use by a wheelchair.

Instruction 2.3

There are some important differences between the three objectives stated above. We would like your group to consider all three statements, determine the differences, and decide which is the best objective.

When you are done, your project leader should read the Feedback.

An engineer should be concerned with all aspects of the system he is designing. He may not do his job properly if he focuses on one facet of the problem and ignores the big picture. Thus, we believe the first objective given in this Instruction is best because it is the broadest statement of the objective of this project. The minute you adopt a narrow objective, such as modifying a set of stairs, you face the danger of eliminating or missing some good solutions. In fact our problem is not to modify the stairs (that is a solution to the problem), but to make it possible for a man in a wheelchair to move safely to and from his work. But before you stop with this objective you should ask yourself, why not try to make this building accessible to anyone who cannot climb stairs? If we can accomplish this goal, we serve other people who have problems as well as the man Mr. Stuart is trying to help.

Instruction 2.4--Information Sources

At this point Tom decided he needed some information about this situation. For example, he figured he would need to know something about the engineering-science principles that might be involved. He would also need to know about the rehabilitation process and what others have done about an accessibility problem. Finally, he would like to know more about the operation of a wheelchair.

What four sources of information is Tom likely to use to find out what he wants to know?

Feedback 2.4--Information Sources

There are four sources of information available to Tom (and to you). One is his own background: although he has not yet graduated, Tom undoubtedly knows a number of engineering-science principles which apply to this situation. In fact, a prime reason for the existence of "schools" is to increase the store of pertinent knowledge so a person has a broad range of concepts and principles readily available to him when he wants to solve a problem. A second source is the literature available: his own engineering books, the library, and the reference material available in the Vocational Rehabilitation office. The third source of information is other people, which might include rehabilitation experts, such as Mr. Stuart, or the man who faces the problem of using a wheelchair, Mr. Masson. Finally, Tom can get information from experimental work. For example, he might want to try out a wheelchair so he can understand what it's like to be in this kind of situation.

Instruction 2.5--Gather Information

Now that Tom has identified potential information sources, he has to decide what information he wants. If you were Tom, what would you like to know; what questions would you want answered?

When you complete your list of questions, your project leader should ask for the feedback. Before he gives you our list, he will compare it with yours to insure that you have not missed any items.

The questions Tom asked were related to the building, the man and wheelchair, and the finances available. The information generated through his discussion with Mr. Stuart is as follows:

1. **The building**

   Mr. Stuart has plans for the first and second floors and a building elevation. These drawings are given on the pages which follow. Note however, that no dimensions are given on these drawings. Your instructor will supply these numbers for your group.

2. **Bob Masson**

   Mr. Masson has made a good adjustment to his handicap, he has full use of his upper body and arms. He can drive himself to work in a car specially equipped with hand controls and can move from the car to the wheelchair and back on his own. The chair is not battery powered, but a new chair might be purchased if that were necessary.

3. **Finances**

   If the company can be convinced that making the building accessible is a desirable goal, they may be willing to help pay for the modifications required. It may also be possible to get State or Federal assistance to modify the building.

Special Instruction

At this point Tom decided he should begin to study some of the engineering-science principles that might be involved in solving this problem. We suggest you do the same by reading the programmed instruction on Conservation of Energy that is provided with this project. Of course you have already learned some basic energy concepts, but there are others you will need to handle this engineering work. Your study begins with material on "The Analysis of a System." This work should be done outside of class while you continue the design process in class.

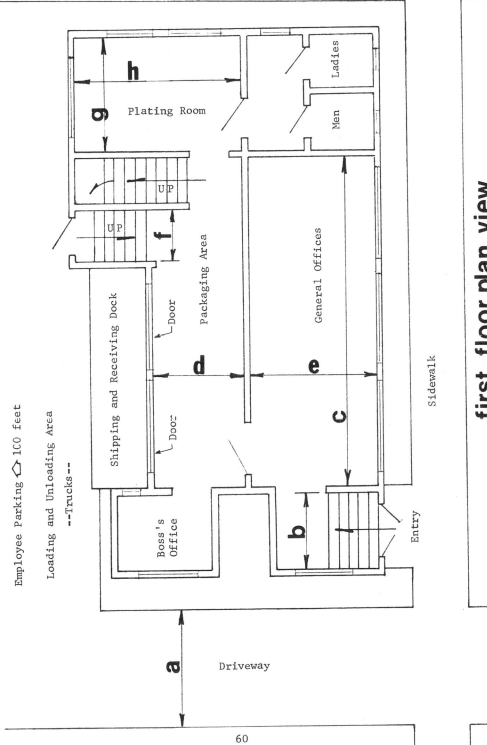

first floor plan view

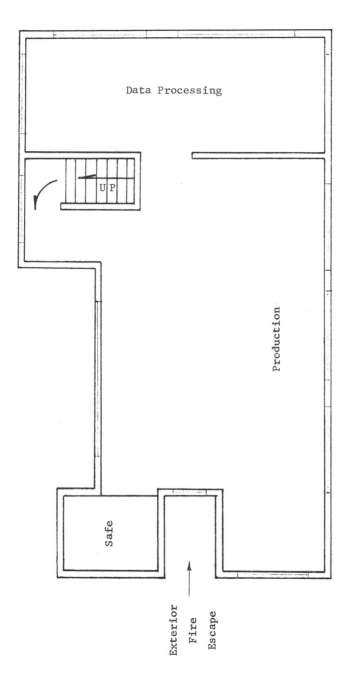

second floor plan view

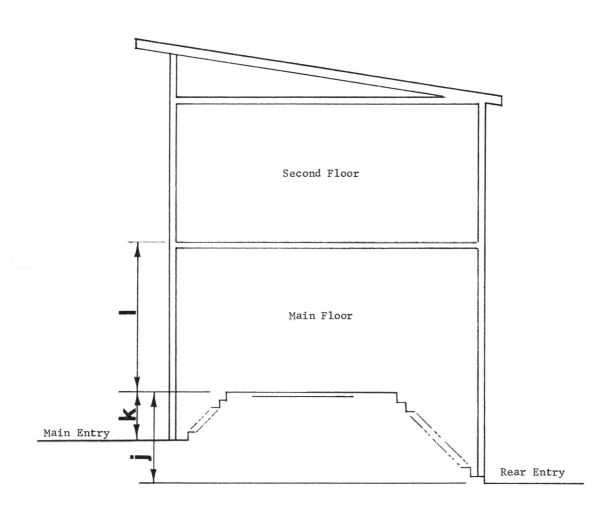

Second Floor

Main Floor

l

k

Main Entry

j

Rear Entry

elevation view

Instruction 2.6--Possible Solutions

Tom and Mr. Stuart now proceeded to generate possible solutions for this project. We suggest your group do the same. List all the solutions developed; do not reject any suggestion at this time. Some ideas which appear unrealistic at first, may well prove to be workable in a modified form later on.

When your list is complete, have a <u>new</u> project leader read the Feedback.

Feedback 2.6--Possible Solutions

Tom's list of possible solutions included the following:

1. Use several people to carry Mr. Masson inside.
2. Build a ramp.
3. Install an elevator using a hand operated pulley-lifting system.
4. Install an electric elevator.
5. Use a fork lift.
6. Move the business to a new building.

Instruction 2.7--Constraints, Assumptions

You may have noticed that Tom has not yet performed the "constraints, assumptions" step in the problem solving process. This was a deliberate decision on Tom's part; he decided he would rather generate solutions first and then use the constraints to help narrow the choice of a solution. This is actually a very common procedure; the constraints step can come either before or after the possible solutions.

What constraints and assumptions are pertinent to this problem? When your list is complete, compare it with the list generated by Tom.

Tom decided that the following <u>constraints</u> would limit what he could do.

1. The building exists and any solution must not interfere with the normal operations of the company.
2. Building codes may limit the changes that are possible.
3. At this point it is not clear who would pay for any modifications or how much money might be available.

Tom also made the following <u>assumptions</u>.

1. The Company is willing to allow some modifications to the building so it is accessible to Mr. Masson and to others. This assumption can be converted to a fact by talking to Mr. Stuart or to the head of the Company.
2. Mr. Masson is willing and able to use the system that is developed.
3. Mr. Masson will only be required to work on the first floor. This assumption can and should be checked and converted into a fact.

Instruction 2.8

What did Tom do next? We would like your group to identify the next step and perform it. We have not identified this step in the process because we want you to do that--part of what you should be learning is how to attack an open-ended problem. Therefore, please identify the next step and perform it. When your work is completed, have a new project leader check the Feedback.

Feedback 2.8--Choose Criteria for Ranking

Tom decided that the next step in the problem solving process was to rank his possible solutions and choose the one or ones to pursue. However, to rank these solutions he had to decide on a set of criteria. The criteria Tom selected are:

1. People--does the solution satisfy the needs of the people involved? In this case the safety of the solution would be of vital importance.

2. Economics--can we afford the solution? Since money is not freely available this project must compete with others for the supply of funds available.

3. Feasibility--can the solution actually be accomplished, does the technology exist or is an extensive research or development program required?

4. Energy Required--if the solution involves manpower, is it possible for a man in a wheelchair to do what is required or will he need assistance?

5. Constraints. After these criteria are applied, Tom still must check his solutions to see that he has not violated any of the constraints.

Instruction 2.9--Rank the Possible Solutions

Tom next decided that he would rank the possible solutions. To do this he assigned a

1. good, average, or poor label for safety.

2. high, medium or low label for cost.

3. yes or no for feasibility.

4. and you can indicate whether manpower or some other form of energy is involved in the operation of the system.

We suggest that your group rank their list of possible solutions in the same way. Then compare your results with Tom's.

Tom's ranked set of possible solutions is given below.

Solution	Safety	Cost	Feasible	Energy
1. Ramp	good	low	yes	manpower
2. Pulley-elevator	medium	low	yes	manpower
3. Electric elevator	good	high	yes	electricity
4. Fork lift	poor	medium	?	diesel
5. Four men	poor	medium	yes	manpower
6. Move building	good	very high	?	---

Tom decided the ramp and the pulley-elevator were the only solutions that appeared to be viable at this time. However, as he proceeds he knows he must check to be sure neither solution will interfere with normal operations or violate any local building codes.

Although the electric elevator is a good solution it was eliminated because of the high expense involved. The fork lift truck might do (if one is available), but there is an obvious safety hazard involved. Asking four men to carry Mr. Masson inside is practical, but it means taking the men away from their work and it involves a potential health hazard for all involved. Tom must also consider the liability roblem if there were an accident. Moving the business is not a very good solution because of the high cost involved.

Instruction 2.10--Analysis

Although we expect your group to follow the general design pattern given for this project, there is no reason why the group must agree exactly with each step. For example, at this point you might choose to consider something other than a ramp and a pulley system. We would encourage you to develop your own ideas and to explore other possibilities. The material which follows shows the pattern of design work for our choice of a possible solution. Most of this can be used as a guide for your work even if you follow a slightly different design path.

We will proceed here on the assumption that the best possible solutions are to use a ramp or a pulley-elevator to make this building accessible to handicapped people. The next stage of the design process usually involves three phases: an analysis phase to determine the significant elements or variables which must be considered, a synthesis phase in which components are designed and organized into a final solution, and an evaluation phase. These three phases frequently occur simultaneously, with earlier decisions being revised as the project develops.

Your group should now begin the analysis phase of their detail design. This might involve two steps, an analysis to identify the basic elements and then a detailed analysis. Start your work by identifying the basic elements of both the ramp and the pulley-lifting system. When the group analysis is finished, the next project leader should read the Feedback.

Tom identified the basic elements of a ramp as:

1. Slope
2. Width
3. Length
4. Room for turns, if required
5. Construction
6. Floor material
7. Side rails

The basic elements of the pulley-lifting system are:

1. Number of pulleys
2. Support for the pulley(s)
3. Loading-unloading system
4. Lifting or lowering mechanism
5. Rope or cable
6. Support for the wheelchair

Instruction 2.11--Detailed Analysis

The point of a detailed analysis is to identify the options that are available and to pinpoint the factors that must be considered. Your group should now perform the detailed analysis for each of these elements. Then compare your work with the analysis prepared by Tom Blaskovics.

Tom's detailed analysis of the ramp is given below.

1. Slope--any slope can be used, but it should be low enough so Mr. Masson can wheel himself up the ramp. A recommended slope is available in the literature.

2. Width--since only one wheelchair at a time is likely to use the ramp it need be only wide enough to accommodate one chair.

3. Length--this is determined by the slope and the space available.

4. Turns--if a turn is required, enough room must be provided so the chair can be maneuvered.

5. Construction--a permanent reinforced concrete ramp or a less permanent wooden ramp might be used.

6. Ramp surface should be safe in rain, snow and ice, if possible.

7. Sides--should provide safety so neither the wheelchair nor people (children) can fall off.

Tom's detailed analysis of the pulley-lifting system is given below.

1. Number of pulleys
    a. one pulley
    b. two pulleys
    c. three or more pulleys

2. Support for the pulley(s)
    a. size and shape: length, diameter
    b. joints
    c. arrangement: attached to the building or free standing
    d. foundation

3. Loading-unloading system
    a. arrangement
    b. safety
    c. alignment with floor

4. Lifting or lowering mechanism
    a. arrangement
    b. safety
    c. locking

5. Rope or cable
    a. size
    b. length
    c. strength

6. Support for the wheelchair
   a. size, shape and weight
   b. load per trip
   c. loading and unloading

Instruction 2.12--Synthesis, Evaluation

At this point your group should be ready to begin the synthesis of the elements required to make this building accessible. During this process you will probably find it necessary to gather a variety of information from one or more of the sources we described earlier, but we believe you can handle both that job and the synthesis-evaluation steps without further prompting from Instruction sheets. Your final design report should present the results of your work with two systems (which might be the ramp and the pulley-elevator or some other combination) and your recommendations to the State Department of Vocational Rehabilitation office. Good luck!

Special Instruction

If the pulley-elevator is one of the systems you are considering, we suggest that you gather information by performing some experimental work with single and multiple-pulley systems. If a wheelchair is available, you might also perform some experiments with this.

The Feedback to this section is your design. Please move on to the next Instruction.

Instruction 2.13--Oral Report, Written Report

At some point your group must make a final decision on the system they recommend to make this building accessible. Before you do, you should carefully check the list of Basic Elements which you prepared earlier to be sure you have not left out any important components of the design. As you work, be alert for further innovations which will make the system safer or easier for handicapped people.

Oral Report.

The specifications you are preparing will be a significant part of your final design report. Completing this report is the step which follows the design specifications. In fact, you probably started preparing your report when you began work on this project, by recording your objective, constraints, etc. However, there is one more checkpoint you should pass before you put too much effort into writing the final draft. This is an oral report to the supervisor of the project (in this case, your instructor and classmates). You should plan to give a ten minute (maximum) oral report on the final design details for the whole project. We would like each member of the group to participate in this oral report by presenting some aspect of your design. However, your group should decide on the number who actually participate. Your instructor (and classmates) can be expected to ask questions of the presentor and the other members of your group during and after the presentation. The whole group, not just the presentor, should be prepared to defend the design decisions. In fact, answering these pointed questions should help you learn how to criticize and evaluate your own work.

The oral report is a commonly used technique in the professional world of the engineer because it requires him to bring together all the details which represent the total design. In addition, it affords an opportunity for others to judge the work and determine if the engineer has developed a completely reasonable, workable and acceptable solution to the original design objective. This final discussion will allow you to look at all aspects of the design and make final modifications in your design specifications.

Written Report.

When you have completed your oral report, each member of the group should be ready to write an individual, final report for the project. If you have kept good

records of your work, you should have most parts of this report completed, including the objective, design constraints, possible solutions, preliminary evaluation, analysis, and synthesis.

One pattern for the report that would be appropriate is given below.

Introduction

Basic Objective

Constraints--Assumptions

Possible Solutions

Evaluation--Decision

Detailed Analysis, Synthesis, Evaluation (which should include a cost estimate)

Final Decisions--Summary

Final Recommendations (including a drawing of the recommended system)

When you have completed your report, turn it in to your instructor. He will give you the Feedback for this section.

The feedback for this section is our version of a Final Project Report.

Before you read our report, read the next Instruction.

Instruction 2.14--Evaluating Your Report

There are several ways in which you can use our report. The first is to evaluate your own report on the basis of

Completeness--Have you reported all the important points?

A second evaluation can be made on the report's

Structure--Is your report developed in a logical manner? It may be similar to ours or be developed in some equally logical alternative pattern.

In addition, we suggest that you also evaluate your personal contribution to the design decisions. Did you contribute a number of ideas? Were they good ideas? Did the group use or follow some of your suggestions?

The result of these evaluations should produce two or three grades and a ranking.

A. Grade your report.

Your instructor may allow you to revise your report until you achieve the grade you wish to earn. This does not mean that you should copy our report, but rather that you should use it as a reference while you develop your own style.

B. Grade your design.

Evaluate your group design. Is your design as good as or better than the other designs which have been suggested?

C. Rank your contribution to the group effort.

Did you contribute more than anyone else, less than all the others, or somewhere in between? This evaluation might be represented by a number between 1 and 5 or 6. Number 1 represents the greatest contribution.

The Feedback to this Instruction will be your project supervisor's acceptance of your report. If he believes the report grade you have given yourself is unrealistic, he may ask you to rewrite your report.

This step completes Design Project II. The Feedback will be the grade you receive on all your work.

===

FINAL REPORT

BUILDING ACCESSIBILITY

By _____

Group _____

Class _____

Date _____

Project Supervisor _____

Introduction

Mr. Chuck Stuart, who works at the State Department of Vocational Rehabilitation office is trying to secure a job for Mr. Robert Masson, who is permanently assigned to a wheelchair as a result of a back injury. A job is available at a local company, but the building is not accessible. I was asked to help devise a solution to this problem.

Basic Objective

To make the building safely accessible not only to Mr. Masson but to all people who cannot climb stairs.

Constraints, Assumptions, Decision Criteria

Constraints.
1. We cannot interfere with the normal operations of the company.
2. Building codes must be observed.
3. Limited funds are available.

Assumptions.
1. The building can be modified if that is necessary.
2. Mr. Masson (and others) will only need to gain access to the first floor.

Criteria for Evaluation.
1. The access must be safe and usable.
2. The access must be low cost.
3. The solution must be technically feasible.
4. If manpower is involved, it must be a reasonable amount.
5. The constraints must be satisfied.

Possible Solutions

Given these constraints, assumptions and criteria we evaluated six possible solutions and decided two were viable. These are:
1. Build a ramp.
2. Install an elevator using a hand operated pulley-lifting system.

Analysis-Synthesis

A detailed analysis was performed to identify the options available and the factors that must be considered. Using this as a base we designed both a ramp and a pulley-elevator system. The detailed design calculations are shown below.

Note: A set of detailed design calculations are not included in this sample report. However, to be complete your solution should probably consider the following factors.

1. Ramp

      Is enough space available to build a ramp at the proper slope and width? Can a man in a wheelchair move up the ramp and open the door to the building? Is a permanent concrete ramp reasonable or should it be made of wood? Does a ramp to the first floor solve the problem?

2. Pulley-Elevator

      Is enough space available? Can a man in a wheelchair use the pulley? Is the system solid and safe? Is it built inside the building or outside? Can a man in a wheelchair move on one side and off another? Can he align the elevator platform with the building floor and lock it in this position? Can he open the door to the building?

Evaluation and Recommendations

A_1. On the basis of our work we recommend the construction of a ramp which is. . . .

or A_2. On the basis of our work we recommend the construction of a pulley-elevator system which is. . . .

B. We also recommend that the inside of the building be modified in the following ways to make it possible for Mr. Masson to move about as he does his work.

    1. The door to the bathroom. . . .

PROJECT III

THE WINFIELD WATER SYSTEM

This project should take about five weeks to complete. It
is coordinated with Programmed Instruction 3 on Fluid Flow.

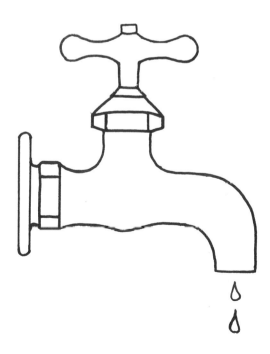

ENGINEERING CONSULTANTS, INCORPORATED

Morgantown, West Virginia

MEMORANDUM

TO: James Smith
FROM: C. A. Arents, Chief Engineer   *Caa*
SUBJECT: Public Water System Design

The town of Winfield, West Virginia, has a water system that was built in 1936. This system has deteriorated to the point where it is no longer dependable. In addition, the Kanawha River, which is the present source of water, is badly polluted and the existing water treatment plant cannot be modified to handle this condition. In fact, since the pollution level is increasing, it seems likely that the river should be abandoned altogether as a source of water.

In view of this situation, the Winfield Town Council applied to the Farmer's Home Administration of the U. S. Department of Agriculture for assistance in financing a new water system. This request has been approved and the Council has authorized our company, Engineering Consultants, Incorporated, to proceed with preliminary plans for the new public water system. If the estimated total project cost is reasonable, the FHA can be expected to provide a grant which will underwrite one-half of the cost of the new system. The town will finance the rest of the cost through a bond sale.

Although your group is composed of engineers who are new to our firm, we think this would be a good project to get you started in this consulting business. Enclosed are four documents we received from the Mayor of Winfield. They provide the preliminary information you need to begin your work. Of course our firm has accumulated a great deal of information on similar projects over the years, so don't hesitate to ask for any data you need to do your work.

PUTNAM COUNTY HEALTH DEPARTMENT

Winfield, West Virginia

The Honorable John Duff
Mayor, City of Winfield
Winfield, West Virginia

Dear Mayor Duff:

Upon reviewing the bacteriological counts on samples of water collected each month from the Kanawha River, this department is of the opinion that a new source of water should be developed immediately. The average bacteria count for eleven samples averaged 17,800 colonies of bacteria/100 milliliters from January through December, 1969.

Chemical composition varies from day to day depending upon the waste discharged by the chemical plants up stream from the treatment plant intake.

Anytime this office may be of assistance, please call at your convenience.

Yours truly,

Gilbert Hall
kew

Gilbert Hall
Sanitarian

GH/kf
cc: Division of Sanitary Engineering

PUTNAM COUNTY HEALTH DEPARTMENT

Winfield, West Virginia

The Honorable John Duff
Mayor, City of Winfield
Winfield, West Virginia

Sir:

Modern sanitation standards depend heavily upon an adequate and desirable water supply. The lack of pressure to flush the commodes at the schools created an undesirable situation recently. Also, should a breakdown of the plant occur, that would require twelve hours or more to repair, there would definitely be a shortage of water for domestic use. Fire fighting equipment could not depend on hydrants.

For years residents have maintained cisterns or other individual water supplies because of the undesirable taste and odor in the public water. Often times I have found containers of water of questionable quality in restaurants because the public supply imparted an undesirable taste in coffee, tea or foods cooked in it.

I urge the Town of Winfield to obtain an adequate and desirable water supply for the health of the school population, residents of the town and to promote the economic development of the town and surrounding area.

Yours truly,

Gilbert Hall
/kew

Gilbert Hall
Sanitarian

GH/kf
cc: W. Va. State Department of Health

TOWN OF WINFIELD, WEST VIRGINIA

PUBLIC WATER SUPPLY

Winfield is the County Seat of Putnam County, West Virginia.

The Town was laid out in 1848 and incorporated in 1868. It was named in honor
of General Winfield Scott, and is located in Scott Magisterial District, situated
by the Kanawha River and along West Virginia Route 17 and connected to Route 35 by
a toll bridge.

The Town is located within easy commuting distance of the Kanawha Valley Industrial complex.

The streets are generally paved.

The utilities are:

      Water System--municipally owned.

      Sewage collection system--municipally owned.

      Telephone service--Chesapeake & Potomac Telephone Company of West Virginia

      Electric power--Appalachian Power Company

      Natural Gas--Winfield Gas Company

Located in the Town:

      County Court House

      County Jail

      County Board of Education

      Federal Agencies

      Newspaper--The Putnam Democrat

TOWN OF WINFIELD, WEST VIRGINIA
PUBLIC WATER SUPPLY

Population Trends

Year	Winfield Population	Population Putnam County
1910	291	18,587
1920	253	17,531
1930	294	16,737
1940	318	19,511
1950	346	21,021
1960	318	23,561
1970	328	
	Plus 8.5% Growth 50 years	Plus 21% Growth 50 years

School Population

School	Elementary	Winfield High
Enrollment	388	621
Teachers	13	19
Others	4	4
ot Lunch Program	x	x
ᴜhowers	no	x
Average Monthly Water Use	34,591 gallons	36,756 gallons

Instruction 3.1--Define the Problem

The first step in the decision-making process is to define the problem. In this case we would like your group to perform a <u>detailed analysis to identify the major factors that should be considered in the design of this water system</u>. For example, one of these factors is a new water treatment plant. Your group should identify the other major factors that should be considered in the design process.

Project Operation.

This project will be organized as a Guided Design--Case Study. This means you will still receive Instructions and Feedback; however, the information we provide will be limited to the identification of the factors involved in a given design step and some of the information required to perform the design work. You will be expected to use this and the other information you gather to complete your design. (Include in your report the name of the source, author, volume, pages, etc. when you use outside information.)

<u>In each step of this project the decisions made by your group will be compared to those made by the Consulting Engineering Firm in the actual Winfield project before you receive the Feedback.</u> For example, in this instruction you will be checked on your ability to identify the major factors in the water system design. The same will be true for each of the other steps you are asked to take. Each one will be checked before you move on.

Now, return to Instruction 3.1 and identify the major factors that should be considered in the design process for a new water system to replace the old deteriorated operation. (Hint: what are the components of a water system and on what basis are they sized?)

Each group must submit a written copy of their decisions to the Chief Engineer (or instructor) before feedback will be provided.

---

eedback 3.1--Define the Problem

In the opinion of the design consultants, the major factors which should be considered in this problem are the following.
A.   Amount of water needed.
B.   Source of water.
C.   Treatment required.
D.   Distribution system.
E.   Pumping required.
F.   Cost of the delivered water.

Note: Neither the words in your list nor the items you chose are likely to be exactly the same as those in our list. Therefore, in this and all the other steps, you may have to convince your design consultant that you have identified an appropriate set of factors when he checks your group's effort.

Programmed Instruction.

The solution of this problem will depend in part on your ability to perform calculations which involve the flow of liquids in a pipeline. To prepare for these calculations, we suggest you begin the study of the programmed material on Fluid Flow that is available with this project.

---

nstruction 3.2--Objective

What is the objective of your work on this project?

Feedback 3.2--Objective

We believe an appropriate objective for this work is:

To design the best possible water system that will satisfy the needs of this community at a reasonable cost.

Instruction 3.3--Water Needs

We assume that your group will solve this problem by working on each of the major factors identified earlier, treating each factor as if it were a sub-objective. At this point we suggest you proceed with detailed work on the first factor we identified, which is

A. The amount of water needed.

To do your work you will require certain information. To get this information you must do two things:

1. Identify the key factors involved in establishing the "quantity of water" Winfield is likely to need; i.e., analyze the Winfield water needs.
2. Describe the information you need to proceed with this part of your work. This information will probably be related to the factors you identify.

As we indicated earlier, your instructor will check to see how well you have done these two tasks; then you will receive the Feedback so you can proceed with the other design steps for this sub-objective. These might include possible solutions, evaluation-choice, analysis, synthesis and the final evaluation.

Feedback 3.3--Water Needs

Analysis.

Our analysis indicates that to calculate the amount of water needed you must consider at least five factors.

1. People
   a. The number of people who will be served by this water system.
   b. The amount of water likely to be used by these people.
2. Fire Protection
   The hydrant system will be part of the water system.
3. Public Buildings
   a. Schools
   b. Court House
   c. Hospital
4. Business and Industry
5. Expected growth and future demand for water (including additional customers outside of Winfield proper who might be served).

Information Needed.

The information you requested might include:

1. The present consumption of water.
2. The expected growth of the town.
   a. People
   b. Industry
3. A plot of the town showing the location of homes, factories and other water users.

Some of this information is provided on the two documents which follow.

Before you start the next Instruction you should provide your instructor with your estimate of the amount of water needed for the new system (i.e., gal/day or gal/month).

TOWN OF WINFIELD, WEST VIRGINIA

ANALYSIS OF WATER SALES

FISCAL YEAR 1967-1968

Month	Gallons Pumped	Average Gallons Per Day	Average Pumping Hours	
July	711320	22945	6.38	
August	790929	25514	7.08	
September	705950	23532	6.54	
October	686836	22156	6.15	
November	646670	21556	6.0	Note: Average pumping
December	653212	21071	5.85	per day for this period
January	667650	21537	5.98	6.32 hours.
February	686940	24534	6.82	
March	637170	20554	5.7	
April	671770	22392	6.22	
May	731420	23594	6.55	
June	708970	23632	6.56	

FISCAL YEAR 1968-1969

Month	Gallons Pumped	Average Gallons Per Day	Average Pumping Hours	
July	732190	23619	6.56	
August	677384	21851	6.07	
September	836990	27899	7.75	
October	771158	24876	6.9	
November	794090	26497	7.36	Note: Average pumping
December	649790	20960	5.82	per day for this period
January	1012376	32657	9.07	7.21 hours.
February	843530	30126	8.36	
March	790690	25506	7.08	
April	772760	24253	7.75	
May	1048770	33831	9.4	
June	933830	31128	8.65	

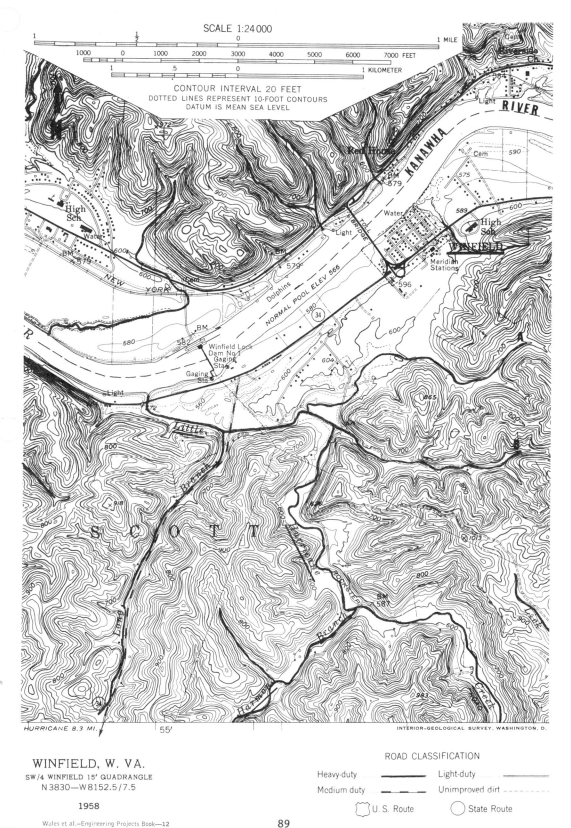

SCALE 1:24 000

CONTOUR INTERVAL 20 FEET
DOTTED LINES REPRESENT 10-FOOT CONTOURS
DATUM IS MEAN SEA LEVEL

WINFIELD, W. VA.

SW/4 WINFIELD 15' QUADRANGLE
N 3830—W 8152.5/7.5

1958

Wales et al.–Engineering Projects Book—12

89

ROAD CLASSIFICATION

Heavy-duty	————	Light-duty	════
Medium duty	▬ ▬ ▬	Unimproved dirt	- - - -

U. S. Route   State Route

Instruction 3.4--Source of Water

Now that you have determined how much water is needed or will be needed, your next step should be to:

1. Generate possible solutions for the source of the water supply.
2. Identify the information you need to perform the design work for this part of the problem.

Feedback 3.4--Source of Water

Possible Solutions.

The potential sources of water appear to be:

1. The Kanawha River--which is polluted, but might be treated in a new plant.
2. Another stream in the area.
3. Wells.
4. Springs.
5. Rain.
6. Another city.
7. A County system or a flood control project, if there is one nearby.

Information Sources.

1. Reports on water sources.
2. Rainfall data.

The next four documents contain some of the information you may need. When your group has finished digesting this information, choose the source of water you expect to use and report this to your design consultant. If he is satisfied with your work, you will be given the next Instruction.

TOWN OF WINFIELD, WEST VIRGINIA
PUBLIC WATER SUPPLY

U. S. ENVIRONMENTAL DATA SERVICE--WINFIELD LOCK STATION

Year	Annual Precipitation		Year	Annual Precipitation	
1931	41.34	inches of	1962	48.03	inches of
1932	41.69	rainfall	1963	52.70	rainfall
1933	43.21		1964	34.86	
1934	35.96		1965	34.18	
1935	57.11		1966	33.00	
1936	38.53		1967	37.10	
1937	44.73		1968	44.48	
1938	47.35				
1939	46.45		1969	(Per Month)	
1940	37.57		Jan.	2.21	
1941	34.65		Feb.	1.18	
1942	40.80		Mar.	1.15	
1943	38.31		Apr.	3.79	
1944	36.06		May	1.62	
945	42.83		June	2.60	
1946	39.78		July	8.62	
1947	35.41		Aug.	6.77	
1948	40.18		Sept.	4.87	
1949	37.23		Total	32.81 (9 Mo.)	
1950	47.72				
1951	44.20				
1952	39.80				
1953	35.16				
1954	37.46				
1955	35.94				
1956	52.84				
1957	34.98				
1958	45.27				
1959	36.20				
1960	32.99				
1961	43.35 (10 Mo.)				

TOWN OF WINFIELD, WEST VIRGINIA

PUBLIC WATER SUPPLY

Precipitation
inches of rainfall

Year	Jan	Feb	Mar	Apr	May	June	July	Aug	Sept	Oct	Nov	Dec	Total
1931	1.48	2.89	4.60	4.59	3.06	2.50	4.82	6.01	4.51	1.75	2.01	3.12	41.34
1941	2.40	0.65	1.59	1.73	1.56	4.80	5.82	5.82	2.68	2.84	2.65	2.11	34.65
1951	4.63	4.16	4.57	3.18	3.46	4.77	2.68	2.35	3.49	1.32	5.02	5.17	44.20
1961	3.90	2.98	4.70	4.52	4.61	5.31	7.14	4.03	1.18	4.98	--	--	43.35 = 10 mo.
1968	3.85	0.70	4.98	2.97	2.97	2.49	3.76	9.10	2.38	4.47	2.10	1.67	44.48

For computing the worse possible condition of run-off volume, use the year 1941.

Summary of a Report for the

TOWN OF WINFIELD, WEST VIRGINIA
PUBLIC WATER SUPPLY

The present source of Raw Water is unsatisfactory. The Town has been instructed by Health Authorities to seek a new source of supply.

The present water treatment plant is in a deteriorated condition.

Proposed Solutions to Problem.

During 1962 the firm of Slonneger, Sheffler, Weaver and Boyle, Incorporated, were employed to prepare a feasibility report to the Town as to a solution to their water problems.

The situations considered were:

I. The use of wells were considered unsatisfactory due to the unavailability of underground water supplies. In addition, the shallow wells have a high mineral content which would require expensive treatment.

II. The use of Little Hurricane Creek was recommended as a source of supply. The flow in this stream is continuous throughout the year. Near the Mouth of the stream is a natural location for a diversion Dam of sufficient holding capacity.

III. The watershed first right going up Little Hurricane Creek (Long Branch) could be used for an impoundment. The run-off area is approximately 175 acres. However, this would require the relocation of the existing road.

IV. The Town can continue to use the water in the Kanawha River. However, the pollution in this river is increasing very rapidly.

Summary of a Report for the

TOWN OF WINFIELD, WEST VIRGINIA
PUBLIC WATER SUPPLY

Recent data collected by the firm of Engineering Consultants, Incorporated.

I. The unavailability of underground water referred to in the 1962 report by Slonneger, Sheffler, Weaver and Boyle was confirmed. The water that is available does have a high mineral content.

II. Due to the expanding developments of the Teays Valley area, Ranch Lake Estates and the normal growth along the improved road extending up Little Hurricane Creek, this source would become polluted within the foreseeable future, rendering this source of supply unusable.

III. The watershed first right going up Little Hurricane Creek (175 acres) was suggested for an impoundment. At this time, the land would be difficult and expensive to obtain. In addition, this supply would have to be supplemented by water from Little Hurricane Creek and this water is subject to pollution.

IV. The Kanawha River is subject to high levels of pollution; thus, it does not merit any serious consideration.

V. An additional source not dealt with in the referenced report of Slonneger, Sheffler, Weaver and Boyle, Incorporated, follows:

Consideration was given to a large flood control program and impoundment proposed by the Corps of Engineers, which would serve a much larger area of Putnam County.

The Town Council met with the Chief Engineer of Engineering Consultants, Incorporated, and discussed the proposed program. The Council's determinations were:

a. The flood control project would be too long range for Winfield to wait.

b. That the cost would undoubtedly be greater for the citizens of Winfield.

Thereupon, the Council asked the Chief Engineer to direct his efforts towards the development of a supply to serve the Town of Winfield and provide for the foreseeable future growth.

Instruction 3.5--Impoundment

The men at Engineering Consultants, Incorporated, decided that an impoundment (reservoir) of rain water would be the best source of supply for Winfield.

Before you can choose a location for this impoundment, you must estimate the watershed acreage required. To do this, you may need some of the following information.

a. The present rate of water consumption is as high as 12×10^6 gal/yr. This might well be doubled in the foreseeable future because of population growth and the increase in per capita use of water that can be expected when a new, clean source is available.

b. Only 40% of the rain and snow which falls on the watershed will actually be collected in the impoundment. The other 60% is lost immediately through percolation into the soil and evaporation. About half of the water which reaches the impoundment can be expected to overflow and be lost during periods of heavy rain; therefore, only 20% of the rain and snow will actually reach and stay in the impoundment.

c. Losses continue to occur after the rain and melted snow reach the impoundment. In addition to the overflow losses mentioned above, there are evaporation and percolation losses. During the hot months these losses from the impoundment account for about 19×10^6 gal/yr.

d. During the winter months there are other losses because some of the snow which falls on the watershed sublimes directly into vapor and never reaches the impoundment. These losses may amount to 10×10^6 gal/yr.

e. It is common practice to store one year's supply of water in an impoundment such as this.

With this information you should be able to determine the size of the watershed needed for the Winfield system.

Before you check the results of your work with your consultant, you should locate both the watershed and the dam you expect to use on the map of the Winfield area.

Feedback 3.5--Impoundment

At Engineering Consultants, Incorporated, they decided that about 300 acres of watershed was required to gather sufficient rainfall for the Winfield water system.

The second left branch (B) off of Hurricane Creek has a watershed of 297 acres, so it should be a good source of supply. (This branch runs almost straight east just below the center of the map that was given earlier. An unimproved road follows the path of the creek.) Note on the Topographic Map that the terrain around this creek is rough and inaccessible, except up the Valley. The ridges are narrow so the land surrounding the watershed is not conducive to population growth. In addition, the neck of the valley is narrow so a dam can be built.

The Topographic Map shows an unimproved road along the ridge, not much different from a Jeep trail. Developers are responsible for roads and streets, neither the County nor State participates in road construction for Developers. The area of usable land is so small no developer could afford access roads in and about the ridge tops for development purposes.

The stream flow from the run-off area is 297 acres. The stream flow was measured by constricted, sharp crest rectangular weir. A series of measurements were made a part of this report and are as follows:

WEIR READINGS proposed source of supply.

24" sharp crest, constricted, rectangular Weir.

Date	Reading	GPM	Gallons Per Day
12/15/69	1-1/4"	98.75	142,200
12/16/69	1-1/8"	84.36	121,478
12/17/69	1-1/8"	84.36	121,478
12/18/69	1-0"	71.76	103,334
12/19/69	0-7/8"	57.46	82,742
12/20/69	0-7/8"	57.46	82,742
12/21/69	0-7/8"	57.46	82,742
12/22/69	0-7/8"	57.46	82,742
12/23/69	0-7/8"	57.46	82,742
12/24/69	1-0"	71.76	103,334
12/26/69	1-1/8"	84.36	121,478
12/27/69	1-1/8"	84.36	121,478

Since you would have great difficulty estimating the cost of an Impoundment--clearing land, dam fill, spillway, drain pipe (to handle overflow), effluent pipe (carries water from the effluent tower to the treatment plant), and effluent tower--we

will give you this information. The cost for this part of the project was estimated to be $50,000.

Design Criteria for Earth Fill Dam.

The Dam for the proposed Winfield Water System will be constructed on the second left Drain to Little Hurricane Creek. Capacity of the impoundment will be approximately 20,500,000 gallons of water from a watershed of approximately 300 acres. The Dam structure will have a crest at elevation 642.5 and a pool elevation of 637.5 allowing 5 feet of freeboard. The width of the crest is 15 feet with each side of the Dam sloping away at a 1 on 3 slope.

The base width of the Dam is maximum at 210 feet. There will be a 16" drain pipe and an 8" effluent pipe through the Dam, each with five (5) concrete water stop collars to prevent impoundment water from following the pipelines under the Dam.

A concrete spillway is used over the Dam to provide for emergency use while the 16" drain will take care of the normal runoff. A selected material clay core placed on solid rock and extending to the Dam's crest to prohibit seepage through the Dam is proposed; the rest of the Dam fill will be of clay, sand, silt, and gravel-mix--all of which will be placed and compacted in 6" layers. On the upstream face of the Dam a 6" layer of broken stone will be placed with the voids filled with smaller crusher run stone and compacted by crawler tracks. The crest and downstream face of the Dam will be seeded as will any areas from which dirt was taken for the dam's construction. Also, all vegetation below the pool elevation will be removed. The Contractor will maintain the dam for one (1) year.

Instruction 3.6-- Treatment-Purification

Even though you may store pure rain water, the water that is finally pumped into the water mains must be treated. This is an appropriate point for your group to gather information. What is common practice in water treatment?

In addition to these specifications, your group should carefully evaluate and determine the GPM* capacity of the new water treatment plant. You may recall that the old plant operated for an average of 7.21 hours. This means someone was on duty during the time the pump operated.

---

*GPM = gallons per minute

Even when a rain water impoundment is involved, the water must be treated before it is fed into the distribution system. It is common practice to use at least:

1. Chemical Treatment--Alum
2. Flocculation
3. Settling
4. Filtration
5. Chlorination
6. Fluoridation

The actual plant chosen for Winfield was a Neptune Aquarius factory-built treatment plant designed to process 200 GPM. Although this is about three times the present GPM rate, the extra capacity is justified by the need to meet expanded use and the need to meet emergency situations, such as a fire.

The operation of a typical water treatment plant is described in reference material available in the library. Since it would not be easy for you to estimate the cost of a plant such as this, we will supply you with the cost estimate developed by Engineering Consultants, Incorporated. They determined that a 200 GPM Water Treatment Plant, appropriate pumps and sludge removal equipment would cost approximately $120,000.

---

nstruction 3.7--Distribution System

Your group should now be ready to begin to plan the water distribution system for the town. However, before you begin, you should perform an analysis to determine the important factors involved in the system. What factors must be considered when you decide how, where and in what to distribute the water to each user.

We believe the following factors are important to the design of a distribution system.

1. Flow in a given pipe, gpm, velocity.
2. Pipe size (diameter).
3. Pressure required at each point.
   a. Maximum pressure allowed.
   b. Minimum pressure likely.
4. Length.
5. Location.
6. Storage of processed water.
7. The cost of pipe and the cost to pump water through that pipe.

Although cost calculations are listed as the last step in this project, there is no reason to wait until that time to get started. In fact you might as well start now. Some of the information you need to estimate the cost of the water distribution system is given on the next three pages. These sheets represent the Bid Schedule submitted by the low bidder on this part of the system. You should note that several items on this Schedule provide for alternate choices. Thus, in item A-1, the bidder chose to use 8" asbestos cement pipe (ACP) installed at $4.50 per linear foot rather than cast iron pipe (CIP). You can determine the total cost f this part of the job by multiplying appropriate numbers and adding the cost of all the pipe, fittings, etc., which are required for the Winfield system.

BID SCHEDULE

TOWN OF WINFIELD

PUTNAM COUNTY, WEST VIRGINIA

IMPROVEMENTS AND EXTENSIONS TO PUBLIC WATER SYSTEM

Contract A

Transmission, Service, and Distribution Lines, Fire Hydrants, and Meters

Note: The contractor's bid price per foot of pipe installed shall include the cost
of trench excavation, road berm maintenance, rock, excavation, pavement
replacement, seeding of lawns or trench area where required, fittings not
included in bid quantities, all thrust blocks, water for flushing, testing,
and initial filling of line.

BID PRICE SHALL INCLUDE SALES TAX AND ALL OTHER APPLICABLE TAXES AND FEES,
INCLUDING STATE ROAD INSPECTION AND PERMITS.

Item	Quantity	Items with Unit Prices Written in Words	Unit Cost	Total Cost
A-1	6,000 lf 1f=linear feet	8" ACP[a], Class 150, Laid	$4.50	
A-1 ALT. 1		8" CTP[b], Class 150, Laid		
A-2		6" ACP, Class 150, Laid		
A-2 ALT. 1	13,000 lf	6" CIP, Class 150, Laid	3.40	
A-2 ALT. 2		6" PVC[c], Class 160, Laid		
A-3		4" ACP, Class 160, Laid		
A-3 ALT. 1		4" CIP, Class 150, Laid		
A-3 ALT. 2	9,000 lf	4" PVC, Class 160, Laid	2.60	

The three types of pipes listed in this table are:

    a. ACP = asbestos cement pipe

    b. CIP = cast iron pipe

    c. PVC = Polyvinyl Chloride pipe

101

Item	Quantity	Items with Unit Prices Written in Words	Unit Cost	Total Cost
A-4	1,270 lf	2" PVC, Class 160, Laid	$ 2.20	
A-5	70 lf	Casing for 8" Pipe, Laid	7.00	
A-6	35 lf	Casing for 8" Pipe, Bored & Jacked	15.00	
A-7	147 lf	Casing for 6" Pipe, Bored & Jacked	14.00	
A-8	216 lf	Casing for 6" Pipe, Laid	6.00	
A-9	137 lf	Casing for 4" Pipe, Bored & Jacked	12.00	
A-10	188 lf	Casing for 4" Pipe, Laid	5.00	
A-11	3	8" Gate Valve w/C.I. Box	200.00	
A-12	11	6" Gate Valve w/C.I. Box	120.00	
A-13	18	4" Gate Valve w/C.I. Box	100.00	
A-14	3	2" Gate Valve w/C.I. Box	90.00	
A-15	12	6" Fire Hydrant Assembly	450.00	
A-16	7	2" Blow-off Assembly	120.00	
A-17	1	Air Relief Valve	130.00	

Item	Quantity	Items with Unit Prices Written in Words	Unit Cost	Total Cost
A-18	45	5/8" x 3/4" Meter Assembly, New	$120.00	
A-19	3,250 lf	3/4" Plastic, Service, Line, Class 160, Laid	1.40	
A-20	600 lf	3/4" Plastic, Service, Line, Class 160, Encased	3.65	
A-21	6	Connection of Existing Water Mains 2" & Larger to New Water Mains	150.00	
A-22	141	Connection of Existing Service Lines Smaller than 2" to New Water Mains	50.00	
A-23	6	Connection of Existing Fire Hydrants to New Water Mains	150.00	

Contract A
Total Base Bid Unit Price
Asbestos Cement Pipe (ACP) $_____

Contract A
Total Base Bid Unit Price
Cast Iron Pipe Alternate (CIP) $_____

Contract A
Total Base Bid Unit Price
(PVC) Polyvinyl Chloride Pipe Alternate $_____

Bidder: _____

Address: _____

By: _____ Date: _____

Instruction 3.8--Water Storage

In case there is a breakdown in the water supply system, clean, stored water must be available to meet the needs of the town, particularly in the event of a fire. Water for these purposes is usually stored in one or more large tanks.

Your job at this point is to determine

1. The amount of water to be stored.

    a. How many tanks should be used in Winfield?

    b. What size tank(s) should be used?

2. Where the tank(s) should be located?

Feedback 3.8--Water Storage

Number and Size.

The original design plans for Winfield called for two water storage tanks of 100,000 gallons each for a total of 200,000 gallons. This would provide the present town with about an 8-day supply of water if everybody conserved as much as possible. However, after all the plans were finalized and the loans had been made, it turned out that the 297 acres of land cost almost double the original price estimate of $14,000. In addition, the design of the dam spillway had to be altered because of an unusual formation of clay at the original site. Since both of these changes resulted in increased costs, the other plans had to be changed to save a corresponding amount of money. The only place such a change could be made was in the storage tanks. Therefore, the design was changed to provide for only one tank with a capacity of 150,000 gallons. Of course two tanks are safer than one, but there was no other choice available in this case.

Location.

Since gravity flow is used to remove water from the storage tank, it must be located at an elevated point. The consultant will check your design to see if you have done this step properly and if the pressure you calculated for service in the town is sufficient.

Instruction 3.9--Optimum Pipe Size

Earlier in this work you were given cost data and the distances involved in installing the pipes for most of this water system. However, there is one other pipeline which must be considered, this is the pipeline from the purification plant to the storage tank. We will assume that this pipeline is 8000 ft long.

We will also assume that between the purification plant, where a 200 GPM pump is located, and the top of the storage tank there is an elevation difference of 150 ft.

Now we want to determine the size of pipe that should be used for this installation. Unless there is some special factor to be considered, such as safety, the two prime factors involved in choosing a pipe size are:

1. The cost of the installed pipe.
2. The cost of pumping the water through the pipe.

The data we presented earlier shows that the cost of the installed pipe is a function of the diameter of the pipe--the larger the pipe diameter, the greater the cost of the installed pipe.

The material you have been studying on fluid flow should make it clear that the diameter of the pipe directly affects the cost of pumping the water--the larger the pipe diameter, the lower the velocity, the lower the friction, and the lower the pumping cost.

One way to balance these two opposing factors is to plot a diagram such as that shown below. Line 1 on this diagram shows that the installed pipe cost increases as the pipe diameter increases. Line 2 shows that the pumping cost decreases as the pipe diameter increases. Line 3 on this graph is the total cost--the sum of points on the other two lines for each pipe diameter.

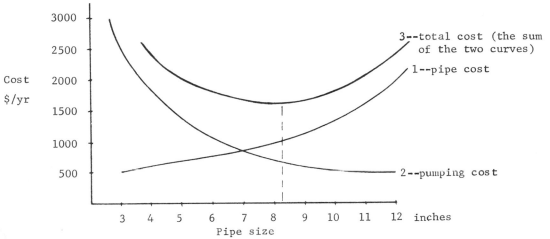

The pipe diameter that lies below the lowest point on line 3, the total cost curve, is called the optimum pipe diameter, because it is the pipe diameter that results in the lowest annual cost for the operation.

To determine the optimum pipe diameter for the 8000 ft long pipeline from the purification plant to the storage tank you will need the following data.

	3"	4"	6"	8" pipe diameter
installed cost	$2.70/1f	$3.00/1f	$3.40/1f	$4.50/1f
lost work, lb_f/sq in/ 100 ft at 200 GPM	3.87	0.985	0.130	0.03
change in elevation	ΔZ = 150 ft in each case			

The cost of electricity usually varies with the amount of energy used. For example, a power company might charge

$0.05 per kw-hr   for the first 30 kw-hrs per month
$0.04 per kw-hr   for the next 150 kw-hrs per month
$0.03 per kw-hr   up to 3000 kw-hrs per month
$0.02 per kw-hr   over 3000 kw-hrs per month

) simplify the work in this project we will assume that a uniform rate of $0.03 per kw-hr applies to all the power used in this pumping process.

The specifications for the pump that will do this job are given below. Actually, for safety reasons two pumps will be purchased and installed. However, only one of these will normally be operated at a given time.

> Deming Vertical Turbine Pump, 8-stage, 200 GPM, 195 ft TDH,*
> 80% efficiency with 20 HP--1750 RPM Motor. Assume Motor
> efficiency of 87.5%.

*TDH = Total Dynamic Head is the feet of water supported by the outlet pressure.

A second set of two pumps, one operating and one spare in case of a breakdown, will be used to move the raw water from the impoundment to the purification plant. The specifications for these pumps are given below. However, you should note that these pumps are not involved in the calculations you are performing in this Instruction.

> Weiman Immersible Pump: 200 GPM, 45 ft TDH,* 58% Efficiency
> with 5 HP--1750 RPM Motor. Assume Motor Efficiency of 87.5%.

The cost of all four of these pumps was included in the cost of the purification plant, which was given earlier.

We suggest you calculate the pumping cost on a yearly basis, $/yr for each pipe diameter. You can assume that on the average, the pump will be operated 8 hours per day, 365 days per year.

Of course the installed pipe cost is only paid once, not every year. To convert this one-time cost to a yearly basis comparible to the pumping cost we will assume that the pipe will last twenty years. Thus, the annual cost of the pipeline is the total installed cost divided by 20 years.

When your group has completed these calculations, plotted their graph, and chosen the optimum pipe diameter for the pipeline from the purification plant to the storage tank, ask your instructor to check the work.

When each group has completed its calculation for the optimum pipe diameter for the pipeline from the purification plant to the processed water storage tank, we suggest you have a whole class discussion to compare and justify your answer.

Instruction 3.10

By now you should have sufficient information to perform a detailed cost analysis for the Winfield water system. This analysis should involve two categories: the capital costs and the operating costs. Perhaps an analogy would be useful in defining these two factors. If you buy a car you either pay cash for the car or borrow part of the money. This is the capital cost. If you buy a used car you might also set aside some money in a contingency fund to buy new tires or to repair some of the components that are likely to fail. If you borrow money, you might also set aside cash for the first interest payment on the capital you borrow.

In addition to the capital costs you must plan for the monthly operating costs. These might include gas, oil, repairs, taxes, insurance, anti-freeze, interest and a payment on your loan--a principal payment.

These same factors apply to the Winfield system. You have been given cost information on several parts of this system: the land, the impoundment, the purification plant, and the piping system. Each of these components is classified as an item of CAPITAL cost. In each case the item must be paid for when the project is built. However, there are other factors which should be included in the capital cost category, two of these are the engineering costs and the legal costs. In addition, the capital cost usually includes a construction contingency--a factor which provides a cushion in case changes are required. This might be 5% of the total of all the other capital costs. In addition, you must provide for the initial interest payment which will be 5% of the amount borrowed.

A second category of costs that must be considered is the OPERATING costs-- the everyday costs that occur after the system is built and operating. We would like you to prepare a list of the factors involved in this category.

When your work is completed, ask your instructor to check your tabulation of the capital costs and the list of the factors involved in the operating costs. When he is satisfied, he will give you some additional data so you can perform the detailed cost calculations.

Feedback 3.10

The Engineering Consultants, Incorporated, cost estimate included the following factors. Note that several items have been left for you to complete.

Capital Investment.

1. Land and Right-of-Way . . . . $ 23,000
2. Distribution System . . . .
3. Treatment-Purification Plant 120,000
4. Storage Tank . . . .
5. Impoundment 50,000
6. Optimum Pipeline . . . .
7. Engineering Costs 44,000
8. Legal Costs 7,500
9. Construction Contingency (5%) . . . .
10. Capitalized Interest (5%) . . . .

Operating Expenses.

1. Wages 5,400
2. Office Expenses 600
3. Taxes, Insurance, Bonds 400
4. Repairs 600
5. Power . . . .
6. Meter reading, billing, collecting 1,300
7. Chemicals 600
8. Interest on Loan (5%) . . . .
9. Principal on Loan ($1,000 first year to $14,000 in the
   40th year) . . . .

Income.

1. Water sales

Your group is now on its own to complete their cost estimate and determine the cost of water to the Winfield customers. What monthly charge do you expect each customer to pay?

In your final report be sure to list the cost of each item given above as well as the total capital cost, the amount of money to be borrowed, the total operating expenses, the monthly income and show how these costs balance; i.e., that the system can be operated at a small profit--which can go into a Reserve Fund for unexpected emergencies.

Your report should also include the detailed calculation for the pipeline from the purification plant to the processed water storage tank.

PROJECT IV

CHOLERA EPIDEMIC

This project should take about five weeks to complete. It is coordinated with Programmed Instruction 4 on Static Forces.

Introduction

You are one of a group of World Health Organization (WHO) consultants that have just arrived in Dacca, Bangladesh. Shortly before you arrived, a typhoon coupled with the normal monsoon rains created a transportation emergency. In addition, there has been an outbreak of cholera. Of course, news of the emergency is on the radio and in all the newspapers, but you are informed about details of WHO's role in the cholera outbreak by a memo from your supervisor.

Instruction 4.1

Read the enclosed memo, discuss the problem and prepare for the meeting by identifying the type of information you believe you will need. When your group has completed this preliminary analysis work, have one man read Feedback 4.1.

WHO

WORLD HEALTH ORGANIZATION

Dacca

Bangladesh

TO: All Consultants

SUBJECT: Cholera Epidemic

FROM: Floyd Harris, Area Supervisor   *FH*

Radio messages from our WHO field workers indicate that cholera is spreading in both the villages where they are working and in nearby rural areas. Vaccination teams are already at work in these villages innoculating the people, but until this operation is completed we must be prepared to treat thousands of people who already have contracted or will contract the disease.

Once cholera is contracted, the best treatment is the injection into or ingestion by patients of large quantities of saline water solution--which replaces lost body fluids.

As you are well aware, the typhoon and resulting flooding have probably disrupted most forms of land travel. Therefore we must take immediate action if our supply of saline water is to reach the infected people in time.

Please give some thought to the problem of distributing the saline water. We will meet tomorrow to discuss and evaluate possible plans.

The first step in any design project is to recognize the problem. The memo presented a problem situation to your group and the group's response, whatever form it took, showed that it recognized the existence of a problem. Now your group should be involved with the process of identifying the information it needs to proceed. If they have not already done so, have your group prepare a list of questions to be used in gathering information about the project. When sufficient questions are prepared, use the information we have given below to answer them. This information is the kind you could obtain from other members of the WHO team that had been stationed in Dacca for some time.

1. Geography

    Your group probably recognized the need for a map of the area. A map has been provided in this section. It shows the essential geographic features of the terrain, including the location of population centers.

2. Transportation: Water--Land--Air

    a. Water--Dacca is built on the delta of the Ganges River. The rivers or streams which form this delta are normally quite shallow, so the usual mode of water transport is by barge or raft. The flooding has caused major changes in the water system which may or may not affect the use of transportation by water.

    b. Land--There is a major highway running from Dacca southwest to the Indian border. Many small roads lead from this highway to the rural centers of population. At the best of times these side roads might be considered barely passable by our standards. However, many bridges and roads have probably been washed out by the typhoon.

    c. Air--Dacca has the only airstrip now in operation. Normally there would be emergency dirt landing strips out in the country, but the flooding has made the use of these secondary strips out of the question.

3. Aircraft

    Several (up to 6) small single engine airplanes are available. Each one can carry a pilot and 900 lb_m of cargo. When fully loaded the planes can cruise at 110 mph for approximately five hours. In addition, there are two helicopters; each has a capacity of 1-1/2 tons and a range of 200 miles at 90 mph.

4. Cholera

    Cholera is caused by Vibrio cholerae bacteria which thrive in tidal estuaries and the human small intestine. Cholera is not very contagious; in general, it is transmitted by food and water and not from person to person if proper hygiene is maintained. Many people develop a natural

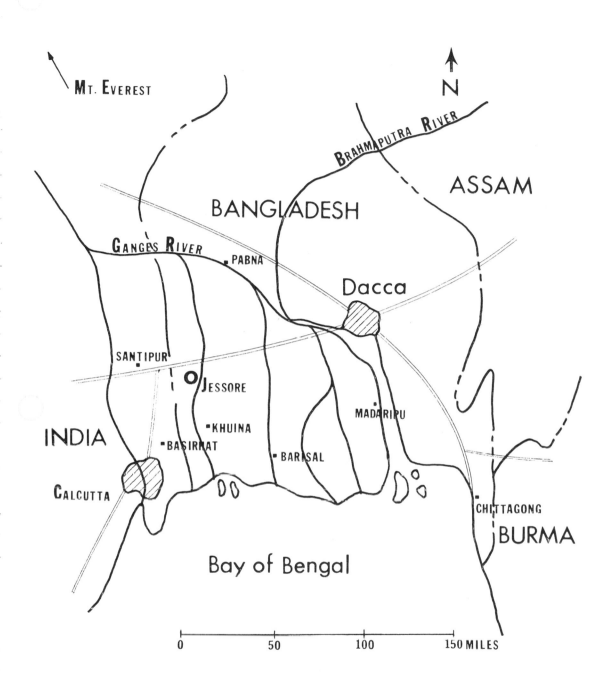

MT. EVEREST

N

BRAHMAPUTRA RIVER

ASSAM

BANGLADESH

GANGES RIVER

PABNA

Dacca

SANTIPUR

O JESSORE

MADARIPU

INDIA

KHUINA

BASIRHAT

BARISAL

CALCUTTA

CHITTAGONG

BURMA

Bay of Bengal

0   50   100   150 MILES

115

immunity to the disease with age through repeated minor exposure to the bacteria. As a consequence about half of the victims of cholera are children who have not yet developed immunity.

The symptoms of cholera usually begin with a feeling of fullness in the abdomen and loss of appetite. Eventually the victim becomes cold, may vomit and then passes great quantities of liquid stool which become more watery as time passes. Painful cramps follow and then possibly deep shock and death. Death comes essentially from the dehydration of the body within hours or days of the onset. Between 50% and 75% of those acutely stricken die if they do not receive prompt and proper treatment.

5. Treatment

If dehydration is the major cause of death then treatment lies in hydration. This can be accomplished by intravenous injections of a special salt-bicarbonate-glucose-water solution which replenishes not only the lost water but salts and bicarbonates as well. Recent developments have shown that after initial correction of hypovolaemic shock with intravenous fluids the treatment can switch to oral replacement of lost fluids. This new method has significant advantages in that as little as 20% of the replacement fluid need be by intravenous injection. Local water sources may be used for oral replacement solutions and the required chemicals should be generally available. As a result of this new method the cost of treatment has been known to decrease from $42 per patient to 63 cents. Although the amount of solution required per patient varies, adults typically require a total of 15 liters and children 7 liters of solution.

6. Saline Water

The saline water solution used in the treatment of cholera consists of 3.5 gm of NaCl, 10.08 gm of $NaHCO_3$, 21.6 gm of glucose dissolved in one liter of water. This solution can be used for both oral and intravenous treatment and is prepackaged in sterilized 1/2 gallon containers similar to the rectangular milk cartons commonly used in the USA. The only precautions necessary for storing this solution is to see that it does not become contaminated; it must remain sterile. Although the present supply of solution is ample, and solution can be prepared in Dacca, a further expansion of the epidemic may severely deplete the supply and overtax production facilities.

7. The Epidemic

Present figures indicate that of the seven million people in the area, there are 10,000 confirmed cases of cholera. One week ago there were 700 cases and one week before that only 40 cases were being treated. The center of the epidemic is around Jessore.

The information given here should answer many of the questions you have concerning this problem. However, this response probably does not answer all of your questions or provide the detailed information you want or need. For example, the map provided here is functional, but more detailed maps exist. Cholera is too complex a disease to describe in a few words. Numbers tend to represent averages or extremes; perhaps more exact values are necessary. Experts do not always agree with each other; several sources should be consulted.

You already know that four basic sources of information are available to you. These are:

1. Your own previous education, training, or experience.
2. Books, other printed matter or media.
3. Other people--experts.
4. Experimental work.

Although the information presented here came from other people, perhaps even experts, there is still much you need to know to help solve this problem. Therefore, now is the time to call on all the resources available. As a start, we suggest you obtain copies of the references on Cholera given below (perhaps you can find other references or local experts).

R. B. Sack, et. al., "The Use of Oral Replacement Solutions in the Treatment of Cholera and Other Severe Diarrhoeal Disorders," Bulletin of the World Health Organization, Vol. 43, No. 3, 1970, pp. 351-360.

Nalin, D.; Cash, R.; and Rahman, M., "Oral (or Nasogastric) Maintenance Therapy for Cholera Patients in All Age-Groups," Bulletin of the World Health Organization, Vol. 43, No. 3, 1970, pp. 361-363.

Hirschborn, N., and Greenough, W. B., "Cholera," Scientific American, Vol. 225, No. 2, August 1971, pp. 15-21.

Pollitzer, R., Cholera, World Health Organization, Geneva, 1959.

Newsweek, "India: A Better Place To Live," June 21, 1971.

Perhaps you can also obtain more information on the geography of Bangladesh, the country, and its people.

When your group's list of questions has been answered, Instruction 4.2 should be read.

Instruction 4.2

While you are in the process of gathering information, one of the other WHO workers asks you to help him with an immediate problem related to loading the vehicles that are transporting the saline solution to the airport and to areas that can be reached by road. As you might expect, a variety of forms of transportation are involved. For example, there are not enough trucks available so even city buses have been pressed into service. This is where the problem occurred.

The 1/2 gallon cartons of saline water solution are stored in the WHO warehouse, packed 16 to a box. The cartons are lying on their sides so leakage will be out of the sterile cartons and can be detected. These boxes are loaded on pallets which are moved out of the warehouse on hand operated dollies. The pallets are lifted by the simple pulley system shown below, which is supported by two 26-ft telephone poles set 16 ft apart.

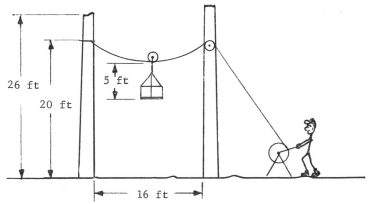

A 1-1/4-inch manila rope is used in this system. During the bus loading operation the part of this rope between the poles has frayed and in one case it broke. We could use a larger rope, but we're not sure this would solve the problem. In addition, it would mean changing the pulleys, which are too small for a larger rope. This problem did not occur when we were loading the trucks. The loaded 4 x 8 ft pallets, which hold 36 cartons, weigh over 2000 lb_f. The buses are 12 ft high and we usually allow a 1 ft clearance during loading. What do you suggest?

Special Instruction

Although you can probably make several "common sense" suggestions to solve this lifting problem, you will be expected to justify your answer with engineering calculations. The analysis of this system will require an understanding of STATIC FORCES. To help you get the appropriate background we suggest you read the programmed instruction that is provided for this project. After you read the first

few parts of this program you should be able to generate several possible solutions to this loading problem and select the best one using what you have learned.

Although each member of your group is expected to read the programmed material at home, you can collaborate on a solution to this problem in class. While you are reading the program on Static Forces, you should continue with your design work in class. Thus, you should now move on to the next Instruction.

Instruction 4.3

From the information you have gathered so far it should be evident that you are involved in a delivery problem. Your solution may involve transportation by water, land and/or air but in any case the packages of saline water appear to be in line for some very rough handling. The worst possible case would be dropping the packages from an airplane. The "Memo" on the next page shows that the WHO director is also thinking along this line.

Your group should carefully consider its response to the request contained in the memo. What do you plan to do? Discuss this before taking action. Then have one member of your group read Feedback 4.3.

WORLD HEALTH ORGANIZATION

Dacca

Bangladesh

TO: All Consultants

SUBJECT: Cholera Epidemic

FROM: Floyd Harris, Area Supervisor   *FH*

Information we have just received makes it quite clear that we must plan to transport the saline water solution to quite a few isolated villages. To make maximum use of the planes at our disposal, we must design and build a device which will hold a 900 lb supply of the solution underneath the plane. The load must be released by the pilot, as he flies low over the drop zone.

Your team has been assigned the job of designing a device to do this job. This work must be completed by tomorrow afternoon so we can construct the system, install it, and begin delivery before the end of the week. The individual parts for this device and the fabrication will be done by local craftsmen. To perform their work they will need a complete set of working drawings.

As project leader for this step, you should be sure your group considers the point of view presented in the following paragraphs.

It appears that your area supervisor has taken this design work into his own hands. He has directed you toward a particular solution before you are ready for this step. This is not an unusual situation and you should be prepared to deal with it in an appropriate way. How you react may well depend on the confidence you have in your own ability. For example, even though you have been educated as an engineer, you may exhibit any one of four "belief systems" identified by some psychologists*. These systems are:

The <u>technician</u> who follows orders and is a dependable employee.

The <u>radical</u> who is against new ideas for the sake of being against something.

The <u>salesman</u> who likes to work with people, hates to be alone, and is influenced by his friends.

The <u>professional</u> who exercises his own judgment, an individualist who is open to diverse inputs, a creative decision-maker.

As you might expect, most people exhibit mixtures of these characteristics and you can probably find good reasons to want some of each type of person in our society. But which type is most appropriate for the educated people who will lead our society? Your response should be a vote for the professional. In fact, the increasing complexity of our society and the magnitude of its problems demands that we develop increasing numbers of these professional people. This is the type of person we hope to develop through the design work you are doing.

If you are operating as a professional, you will probably respond to the memo from Floyd Harris by meeting with him, explaining that his input is worthwhile, but premature, and indicate that you expect to reach a decision shortly. Such candor should ease the situation and allow for progress toward the best possible solution.

When your group completes their discussion of these ideas, continue with the next Instruction.

---

*<u>The Affective Domain</u>, Contributions of Behavioral Science to Instructional Technology-1, Harvey, O. J., pp. 67-96, Communication Service Corporation, Washington, D. C., 1970.

If your group wanted to present Mr. Harris with some evidence related to his suggestion, you might simulate the air-drop of a 1/2 gallon carton of water. What is the likelihood that a carton would survive such a drop even if the plane could fly very low?

You might also do a preliminary evaluation of his possible solution by asking some "what happens if" type of questions. For example, suppose your group did decide to use the airplanes to airdrop the saline solution. Explore this idea and see if you can agree on the probable outcome of this course of action. Even if you could drop a 900 lb_m load successfully, what is likely to happen on the ground?

When your group has completed their discussion, read Instruction 4.5. The Feedback to this step will follow this pair of Instructions.

nstruction 4.5

Since the available airplanes have a very limited capacity, the amount of saline solution delivered on each trip will be small. For example, in some cases it might be possible to airdrop enough vaccine for only 10% of the people who are ill in a given village. What do you expect the result will be? What would happen if the comparable situation occurred in your area?

If you airdrop only enough saline solution for 10% of those who are ill, the result will likely be a riot in which many people are killed or injured. In fact, the riot may be more devastating than the cholera epidemic.

In addition, the capacity of the available airplanes is so small, that you must surely play "God" in deciding which village will receive saline solution first and which will never hear the drone of the airplane engine.

Instruction 4.6

Although you now have some good arguments against Mr. Harris' proposal, you should not eliminate the use of the airplanes, they may still play some part in the possible solutions that might be used. However, before this decision is made you should get back on track with the design process.

You have gathered considerable information and should be ready for the next step in this design project. The group should determine what this step is, then perform it. (We have not identified the next step by name because by now you should know the recommended steps in a design project. We will ask you to select and perform the next step in this way throughout this project.) When your work on the next step has commenced, have a new project leader read the Feedback.

Feedback 4.6--Basic Objective

If your group is following the recommended design procedure, they should now be completing their statement of the Basic Objective for this project. If they are not doing this, we suggest they do it now. It is extremely important that you state the Basic Objective as early in the design process as possible. Without such a statement you cannot focus your thoughts on the problem to be solved.

The Basic Objective you write must state the _real_ problem to be solved. It should be a statement which is broad enough so no reasonable possible solution is eliminated. The following paragraphs illustrate the process we used to move from some limited objectives to a much broader objective. We hope you can use this or a similar process to direct your group to the broadest possible objective.

As we went through the process of generating our basic objective, we decided there were three different ideas we might pursue. These were to

1. deliver the saline water solution.
2. develop a plan to treat the cholera victims.
3. prepare the saline water solution for delivery.

On further consideration of the first memo from Floyd Harris and from the information we gathered, it became evident that we could safely _assume_

1. other people would handle the treatment of the victims if only we would deliver the solution.
2. the saline water solution was already available in Dacca and in sufficient quantities, at least for now.

We then felt prepared to state a basic objective that achieved the desired result, but did not rule out any reasonable solution.

When your group has stated their basic objective, ask them to read this page. Then _continue with your development of a basic objective_ by reading the next Instruction.

Instruction 4.7--Basic Objective

Your basic objective may take one of many forms. It might be as simple as the first one we developed,

> Deliver saline water solution to the cholera areas.

or as complex as

> Develop a system to ensure that saline water solution is available when and where it is needed.

We considered both of these statements in developing our objective and had trouble deciding between them. One, as you can see, is short and to the point; the other adds an important element to the project.

Consider these two objectives along with your objective. What extra element is included in our second objective? Is this an attitude an engineer should have?

After an appropriate period of discussion, have one of your group read the next Feedback.

==

We finally settled on the second objective:

> Develop a system to ensure that saline water solution is available when
> and where it is required.

Although this objective does not specifically use the word "deliver" (and that is the major action involved in this project) it is implied in the statement. But there is also the implication that if the present supplies of saline water run out we will arrange for other supplies, even if we have to make our own. And at the other end of the delivery system, if necessary, we will ensure that the solution reaches the patients. It is not that we wish to get into the business of producing saline water solution, nor do we wish to take over the roles of the doctors and nurses--but we do feel responsible for the success of the whole operation. We cannot accept only the narrow role of a delivery man.

As the project leader for this step, we hope you can ask questions which will lead your group to consider the idea that this total-responsibility attitude is something the engineer should accept. To live up to this responsibility the engineer must be a capable decision-maker, he must be able to work with other people who have different levels of education, and he must deal with a variety of beliefs. For example, the engineer may work with the technician who will make the device he designed. He may also have to deal with a radical village leader who does not want his people to take the saline solution and the WHO field team that is trying to convince the people they should accept the help available. And he must work with different professionals, the doctor, the researcher, administrators, and other engineers.

When the group finally decides that their objective is broad enough and that the level of responsibility they wish to accept is written into the basic objective, they should proceed to the next Instruction.

Instruction 4.8--Constraints and Assumptions

One of the primary reasons for asking questions at the beginning of a project is to gather information about the constraints and assumptions that apply.

Constraints are factors which cannot or must not be changed, and which are likely to influence the choice of a possible solution. Thus, the fact that a day has 24 hours is not listed as a constraint because this is not likely to affect the choice of a solution. In contrast, the fact that daylight lasts only 10 hours might be listed as a constraint in a project which required natural light.

Assumptions are used to simplify the problem so it can be solved or solved with greater ease. For example, we have already made two assumptions:

1. that other people would handle the treatment of the cholera victims.
2. that saline water solution is available in Dacca and in sufficient quantities, at least for now.

There are numerous other assumptions you might list. But just as with constraints you must use your judgment. There is no point in listing every piece of information that is available.

Your group should now analyze the problem and generate a list of the constraints and assumptions which apply to this project. When your list is complete, have one member of your group read the Feedback.

Feedback 4.8

To illustrate the selection of <u>constraints</u>, let's consider the information given in this project in light of the basic objective. In essence, the objective indicates that a delivery problem must be solved. Thus, let us first look at limitations in the transportation system. What is evident from the information given is that

> Surface transportation is normally slow.
>
> Surface transportation is disrupted by flooding at the present time.
>
> Air transportation is possible but Dacca has the only airstrip in
>> operation.

Your <u>assumptions</u> may include the two mentioned in this Instruction plus the following.

> A communication system exists. This will allow us to determine the
>> specific needs for saline water solution at a given location.

Your list of constraints and assumptions will probably evolve and change as you go through the design process right up to the final report. In fact, you should expect to identify new factors as you generate possible solutions or identify sub-objectives. However, the list you present in the final report should be polished and compressed so that it conveys all the pertinent design factors in the fewest possible words. The final list need not tell the reader about all the deadends you considered in preparing the design.

When your group completes its work on the constraints and assumptions, they should proceed to the next step in the design project. The group should determine what this step is, then perform it. When your work is complete, have a new project leader read the next Instruction.

Instruction 4.9--Possible Solutions

The step your group should be performing now is the preparation of a list of possible solutions for an objective similar to this.

> Develop a system to ensure that saline water solution is available when and where it is required.

This step will involve both analysis and synthesis, but the evaluation step should be postponed until after the generation process is completed.

If the group is not preparing this list, ask them to begin this work now. When their work is finished, have one member of the group read the Feedback.

When your group's list is complete, compare it to the list we have provided. You may have solutions we did not think of--that's good. If any of our solutions were not discovered by your group, lead them in discussion until they generate these solutions for themselves.

Possible Solutions.

1. Bring those with cholera to Dacca for treatment.
2. Airlift the solution to the villages using airplanes.
3. Airlift the solution to the villages using helicopters.
4. Ship saline solution by surface transportation.
5. Isolate the affected areas and let the epidemic run its course.
6. Institute measures to prevent future epidemics.

In addition to the specific solutions given here, you may have listed numerous combinations of these possible solutions. For instance, we would also consider the following.

7. Send the saline solution to accessible villages by surface transportation and use an airlift only for isolated villages.
8. Airlift the saline solution to key villages which serve as distribution points for the surrounding villages using surface transportation.

When your group completes its work on this stage of the design, they should identify and proceed to work on the next stage of the project. This stage requires a great deal of involved analysis so we suggest you divide this work among your group, assigning one part of the task to each individual, rather than having the whole group work together on all the tasks. Then have the whole group discuss the individual results.

When the group has begun this new task, have a new project leader read the next Instruction.

131

Your group should probably perform their preliminary evaluation of the possible solutions with the aid of Design Decision Tables. If they did not use the Tables you might suggest that they do so now. Divide the possible solutions to be evaluated between the members of the group. It is not necessary for everyone to evaluate every possible solution. Don't forget to fill out the explanation columns as the solutions are judged acceptable or unacceptable with respect to each factor. It is this explanatory information which you will probably need in undertaking the next stage in the design.

When all the Design Decision Tables are completed, the group should discuss and rank the possible solutions from the best to the poorest. We have intentionally provided no feedback to this instruction because we do not wish to unduly affect your decision in the next design step. When you have completed this ranking you should be prepared for the meeting with Mr. Harris.

DESIGN DECISION TABLE

Objective _____

Possible Solution _____

Factor	Acceptable	Unacceptable	Explanation
1. Does the solution satisfy the basic objective of the project?			
2. Is the solution theoretically feasible?			
3. Is the solution practical?			
4. Does the solution satisfy all the social and ecological factors involved?			
5. Is the effort (time, energy, money) invested in the project worth the value received?			
6. Can the project be completed in a reasonable time?			
7. Is the solution legal?			
8. Are there any special side benefits produced by the solution?			

133

Instruction 4.11

At the start of the staff meeting Mr. Harris announced that as a result of your earlier inputs about the limitations of the light planes he had appealed to the nations of the world for a supply of helicopters. The initial appeal had already been answered and several were on their way--they were expected to arrive in two days. As soon as these helicopters arrive, the group must be prepared to load and deliver the saline solution.

These new helicopters are similar to those already on hand, but <u>they have a cargo capacity of 4 tons</u>. Some are equipped with a side mounted hoist which has a capacity of 2 tons. This hoist is mounted directly above a 4 ft wide by 7 ft high door. On others the entire 4 ton load can be suspended below the craft from a single centrally attached hook and hoist.

One result of the discussion at the staff meeting was the suggestion that your group perform two jobs related to the loading of these helicopters.

1. It was suggested that the WHO team could load one or more 4 x 8 ft pallets on the new side door-helicopters. The single cable on the hoist would be attached to a loop sling, with one part of the rope passing over each side of the load, as shown below. This is the pattern used now to load the trucks and buses.

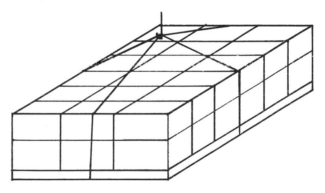

The group decided that it would be desirable to increase the size of this load by stacking the boxes four high. However, they are concerned that each 1-1/4-inch manila rope may crush the top cartons it passes over. Therefore, they want your group to perform the necessary calculations and experimental work to check this out. We do not have the time or the equipment to test the complete unit so you'll have to work with one 1/2-gallon carton.

2. The staff also suggested that on the other helicopters the load could be carried in a 4 x 8 ft scoop-type of box supported by three manila ropes attached to the single cable underneath the helicopter.

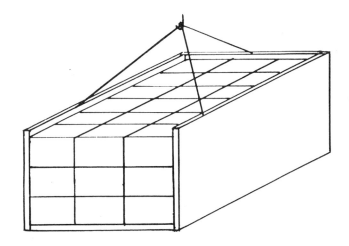

With this arrangement the pilot can either set the box on something so it will tip and allow the load to slide off or an extra rope can be attached to the rear of the box so that end can be lifted. In this case it was suggested that your group determine the size of the ropes needed so we can have these "scoops" ready when the helicopters arrive.

The imminent arrival of the helicopters appears to fix the best possible solution for the delivery system. The suggestions for the package design should be pursued, but that doesn't mean you are bound by either idea. If you believe there is a better way, you should explore that too. In any case, the work you do will be the detailed analysis, synthesis and/or evaluation for this part of the problem.

There is no Feedback to this assignment. The results of this work will constitute the main part of your final design report.

Instruction 4.12

Earlier in this project we suggested that the engineer should accept responsibility for all aspects of the problem he is trying to solve. If you agree with
this concept you should recognize the need to perform a detailed analysis of the
whole delivery system. Of course the choice of the helicopter and the design of
the package are two of the factors that should be considered during this phase of
the work. But what other factors should be included in your plans? To answer
this question your group should determine all of the detailed components involved
in your distribution plan. When this analysis is complete, have a new project
leader read the Feedback.

Feedback 4.12

Your group should have completed a detailed analysis such as that shown below. If any items are missing from your list, ask questions which will lead the group to complete their list.

Distribution Plan--Detailed Analysis

1. Amount and type of solution or salts available and required.
    a. Location of preparation plant.
    b. Location of distribution point for airlift.
2. Storage and handling at distribution point.
3. Distribution pattern.
    a. To each village.
    b. To central locations, then
        1) Deliver by land.
        2) Deliver by water.
4. Airlift Plan.
    a. Delivery order-flight plan.
    b. Role of helicopters.
    c. Role of airplanes.
5. Communications.
    a. To determine the amount of solution required.
    b. To coordinate distribution.
    c. To confirm successful delivery.

You are now on your own to complete this project. However, before you proceed we would like to make one other point. To solve this problem or any problem your group should know how to use each of the steps in the decision-making process. As you work you should note the importance of three of these steps: analysis, synthesis and evaluation. Although these three steps are listed in the process itself, they are also involved in each of the other steps. For example, to gather information you must analyze your needs, determine the best source of material; i.e., synthesize a course of action, and evaluate the information you obtain. To identify the correct problem you must analyze the situation before you and evaluate every possible aspect. The act of generating possible solutions may involve a simple analysis, if the situation is one that is familiar to you. However, if the situation is a completely new one, the generation step may involve a very high level of synthesis. These examples should make it clear that analysis, synthesis and evaluation are critical steps in the decision-making process so you should be concerned with how well you can handle each of them.

The result of your work should be a written report which describes all of the decisions you have made. Each group should also be prepared to make an oral report on their ideas the week before the written report is due.

---

†

HETERICK MEMORIAL LIBRARY

3 5111 00369 7046